Written according to new CBCS (Choice Based Credit System) syllabus of UGC (2017 onwards) and all Indian Universities

POLYMER CHEMISTRY

Beginners to Researchers

SEM-VI

University of Calcutta ■ Jadavpur University ■ Presidency University (UG, Sem-V+VI)
West Bengal State University ■ University of Burdwan ■ University of Kalyani (UG, Sem-V)
Vidyasagar University (UG, Sem-VI + PG, Sem-IV) ■ University of North Bengal
Bankura University ■ Sidho-Kanho-Birsha University
University of Gour Banga (UG, Sem-V) ■ Cooch Behar Panchanan Barma University
Kazi Nazrul University (PG, Sem-III) ■ Visva-Bharati University (PG, Sem-I+IV)
Diamond Harbour Women's University (PG, Sem-III) ■ All Indian Universities

Dr. Ashesh Garai

Head, Department of Chemistry, Rammohan College, Kolkata;
Ex-researcher in Indian Association for the Cultivation of Science (India),
National Institute for Materials Science (Japan),
University of Liverpool (United Kingdom),
University of Southampton (United Kingdom).

Published by :

Santra Publication Pvt. Ltd.

15, Shyamacharan Dey Street,
Kolkata-700073

First Edition
April, 2021

ISBN : 978-81-948683-4-7

Printed by
Abhinaba Mudrani, Kolkata

DEDICATION

The book is

dedicated to

*my wife, **Kakali***

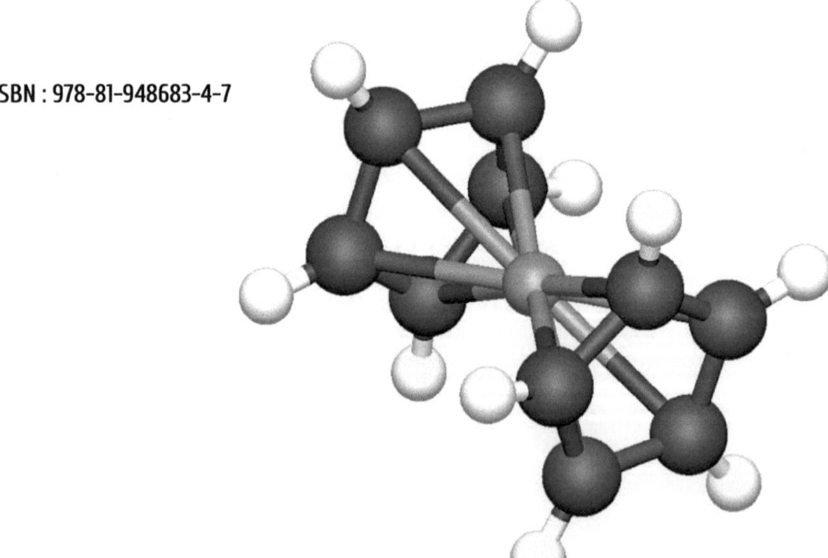

The book on **Polymer Chemistry** is written following the latest UGC recommended CBCS (Choice Based Credit System) syllabus for UG (Hons.) and PG Courses of **all Indian Universities**. Our objective is to present a detailed outlook of **Polymer Chemistry** syllabus of under Graduate & PG courses incorporated by the various Indian Universities.

The purpose of this book is to familiarize undergraduate students of chemistry CBCS programs and entry-level researchers in the polymer science field. The author has written this book in a compact way using various concepts, theories, schemes and images covering history of polymeric materials, functionality, kinetics of polymerization, crystallization and crystallinity, molecular weight and glass transition temperature of polymers, structure and properties of polymers, and polymer solution topics. The book contains 9 Chapters viz. History and Classification of Polymers, Functionality and its importance, Kinetics of polymerization, Crystallization and Crystallinity, nature and structure of Polymers, molecular weight of Polymers, glass transition, temperature of Polymer, Polymer solution, Preparation, structure, properties and application of Polymers.

The author would like to thank Prof. Arun Kumar Nandi, Prof. Izumi Ichinose and Prof. Darren Bradshaw for their guidance in research and also in personal life. He would like to thanks his seniors, colleagues and juniors, whom he has met since he started his doctorate journey. He also would like to thank all his family members especially his two little sons Arpik and Ayush for their support and encouragement.

In spite of the sincere efforts that have been put, there may be some misprints and some inadvertent mistakes. We promise that those will be corrected in the next edition. In addition, any kind of constructive suggestions and comments via email *contactus@santrapub.com* will be highly appreciated.

April 01, 2021   **Ashesh Garai**
Kolkata

SYLLABUS

SEMESTER-VI

CHEMISTRY-C XI : ORGANIC CHEMISTRY-IV

Polymers :   *(8 Lectures)*

- Introduction and classification including di-block, tri-block and amphiphilic polymers; Number average molecular weight, Weight average molecular weight, Degree of polymerization, Polydispersity Index.

- Polymerisation reactions-Addition and condensation -Mechanism of cationic, anionic and free radical addition polymerization; Metallocene-based Ziegler-Natta polymerization of alkenes; Preparation and applications of plastics – thermosetting (phenol-formaldehyde, Polyurethanes) and thermosoftening (PVC, polythene).

- Fabrics – natural and synthetic (acrylic, polyamido, polyester); Rubbers – natural and synthetic: Buna-S, Chloroprene and Neoprene; Vulcanization; Polymer additives; Introduction to liquid crystal polymers; Biodegradable and conducting polymers with examples.

CHEMISTRY-DSE : POLYMER CHEMISTRY

(Credits : Theory-06, Practicals-02)

Theory : 60 Lectures

Introduction and history of polymeric materials :   *(4 Lectures)*

- Different schemes of classification of polymers, Polymer nomenclature, Molecular forces and chemical bonding in polymers, Texture of Polymers.

Functionality and its importance :

(8 Lectures)

- Criteria for synthetic polymer formation, classification of polymerization processes, Relationships between functionality, extent of reaction and degree of polymerization. Bifunctional systems, Poly-functional systems.

Kinetics of Polymerization :   *(8 Lectures)*

- Mechanism and kinetics of step growth, radical chain growth, ionic chain (both cationic and anionic) and coordination polymerizations, Mechism and kinetics of copolymerization, polymerization techniques.

Crystallization and crystallinity : *(4 Lectures)*

- Determination of crystalline melting point and degree of crystallinity, Morphology of crystalline polymers, Factors affecting crystalline melting point.

Nature and structure of polymers : *(2 Lectures)*

- Structure Property relationships.

Determination of molecular weight of polymers :   *(8 Lectures)*

- (M_n, M_w, etc) by end group analysis, viscometry, light scattering and osmotic pressure methods. Molecular weight distribution and its significance.

- Polydispersity index.

Glass transition temperature (T_g) and determination of T_g :   *(8 Lectures)*

- Free volume theory, WLF equation, Factors affecting glass transition temperature (T_g).

Polymer Solution :   *(8 Lectures)*

- Criteria for polymer solubility, Solubility parameter, Thermodynamics of polymer solutions, entropy, enthalpy, and free energy change of mixing of polymers solutions, Flory-Huggins theory, Lower and Upper critical solution temperatures.

Properties of Polymers :   *(10 Lectures)*
(Physical, thermal, Flow & Mechanical Properties)

- Brief introduction to preparation, structure, properties and application of the following polymers : polyolefins, polystyrene and styrene copolymers, poly(vinyl chloride) and related polymers, poly(vinyl acetate) and related polymers, acrylic polymers, fluoro polymers, polyamides and related polymers. Phenol formaldehyde resins (Bakelite, Novalac), polyurethanes, silicone polymers, polydienes, Polycarbonates, Conducting Polymers, [polyacetylene, polyaniline, poly (p-phenylene sulphide polypyrrole, polythiophene)].

UNIVERSITY OF CALCUTTA

SEMESTER-VI

DSE-B-3 : POLYMER CHEMISTRY

(Credits : Theory-04, Practicals-02)

Theory : 60 Lectures

Introduction and history of polymeric materials :   *(04 Lectures)*

- Different schemes of classification of polymers,

Polymer nomenclature, Molecular forces and chemical bonding in polymers, Texture of Polymers.

Functionality and its importance :
(08 Lectures)

- Criteria for synthetic polymer formation, classification of polymerization processes, Relationships between functionality, extent of reaction and degree of polymerization. Bi-functional systems, Poly-functional systems.

Kinetics of Polymerization : *(08 Lectures)*

- Mechanism and kinetics of step growth, radical chain growth, ionic chain (both cationic and anionic) and coordination polymerizations, Mechanism and kinetics of copolymerization, polymerization techniques.

Crystallization and crystallinity :
(04 Lectures)

- Determination of crystalline melting point and degree of crystallinity, Morphology of crystalline polymers, Factors affecting crystalline melting point.

Nature and structure of polymers :
(04 Lectures)

- Structure Property relationships.

Determination of molecular weight of polymers :
(08 Lectures)

- (M_n, M_w, etc) by end group analysis, viscometry, light scattering and osmotic pressure methods. Molecular weight distribution and its significance.Polydispersity index.

Glass transition temperature (T_g) and determination of T_g :
(08 Lectures)

- Free volume theory, WLF equation, Factors affecting glass transition temperature (T_g).

Polymer Solution : *(08 Lectures)*

- Criteria for polymer solubility, Solubility parameter, Thermodynamics of polymer solutions, entropy, enthalpy, and free energy change of mixing of polymers solutions, Flory- Huggins theory, Lower and Upper critical solution temperatures.

Properties of Polymer : *(08 Lectures)*
(Physical, thermal, Flow & Mechanical Properties)

- Brief introduction to preparation, structure, properties and application of the following polymers : polyolefins, polystyrene and styrene copolymers, poly(vinyl chloride) and related polymers, poly(vinyl acetate) and related polymers, acrylic polymers, fluoro polymers, polyamides and related polymers. Phenol formaldehyde resins (Bakelite, Novalac), polyurethanes, silicone polymers, polydienes, Polycarbonates, Conducting Polymers, [polyacetylene, polyaniline, poly (p-phenylene sulphide polypyrrole, polythiophene)].

PRESIDENCY UNIVERSITY

SEMESTER-V

2. CHEMISTRY-DSE : POLYMER CHEMISTRY
(Credits : Theory-06)
60 Lectures
Full Marks : 70

Introduction and history of polymeric materials :
(4 Lectures)

- Different schemes of classification of polymers, Polymer nomenclature, Molecular forces and chemical bonding in polymers, Texture of Polymers.

Functionality and its importance :
(8 Lectures)

- Criteria for synthetic polymer formation, classification of polymerization processes, Relationships between functionality, extent of reaction and degree of polymerization. Bifunctional systems, Poly-functional systems.

Kinetics of Polymerization : *(8 Lectures)*

- Mechanism and kinetics of step growth, radical chain growth, ionic chain (both cationic and anionic) and coordination polymerizations, Mechanism and kinetics of copolymerization, polymerization techniques.

Crystallization and crystallinity :*(4 Lectures)*

- Determination of crystalline melting point and degree of crystallinity, Morphology of crystalline polymers, Factors affecting crystalline melting point.

Nature and structure of polymers :
(2 Lectures)

- Structure Property relationships.

Determination of molecular weight of polymers :
(8 Lectures)

- (M_n, M_w, etc) by end group analysis, viscometry, light scattering and osmotic pressure methods. Molecular weight distribution and its significance. Polydispersity index.

Glass transition temperature (T_g) and determination of T_g : *(8 Lectures)*
- Free volume theory, WLF equation, Factors affecting glass transition temperature (T_g).

Polymer Solution : *(8 Lectures)*
- Criteria for polymer solubility, Solubility parameter, Thermodynamics of polymer solutions, entropy, enthalpy, and free energy change of mixing of polymers solutions, Flory-Huggins theory, Lower and Upper critical solution temperatures.

Properties of Polymers : *(10 Lectures)*
(Physical, thermal, Flow & Mechanical Properties)
- Brief introduction to preparation, structure, properties and application of the following polymers : polyolefins, polystyrene and styrene copolymers, poly(vinyl chloride) and related polymers, poly(vinyl acetate) and related polymers, acrylic polymers, fluoro polymers, polyamides and related polymers. Phenol formaldehyde resins (Bakelite, Novalac), polyurethanes, silicone polymers, polydienes. Polycarbonates, Conducting Polymers, [polyacetylene, polyaniline, poly (p-phenylene sulphide polypyrrole, polythiophene)].

SEMESTER-VI

CHEM06C14 : ORGANIC CHEMISTRY-V

(Credits : Theory-06)

60 Lectures

Full Marks : 70

Polymers : *(12 Lectures)*
- Introduction and classification including di-block, tri-block and amphiphilic polymers; Number average molecular weight, Weight average molecular weight, Degree of polymerization, Polydispersity Index.
- Polymerisation reactions -Addition and condensation -Mechanism of cationic, anionic and free radical addition polymerization; Metallocene-based Ziegler-Natta polymerisation of alkenes; Preparation and applications of plastics – thermosetting (phenol-formaldehyde, Polyurethanes) and thermosoftening (PVC, polythene).
- Fabrics – natural and synthetic (acrylic, polyamido, polyester); Rubbers – natural and synthetic: Buna-S, Chloroprene and Neoprene; Vulcanization; Polymer additives; Introduction to liquid crystal polymers; Biodegradable and conducting polymers with examples.

JADAVPUR UNIVERSITY

SEMESTER-VI

Paper : DSE/Chem/TH/04 : (Choice Based Paper)

4 Credits

Full Marks : 50

4 L/W   60L

Unit : 6042-0 : Polymer Chemistry
(60 Lectures)

1. Introduction and history of polymeric materials : *(4L)*
- Different schemes of classification of polymers, Polymer nomenclature, Molecular forces and chemical bonding in polymers, Texture of Polymers.

2. Functionality and its importance : *(6L)*
- Criteria for synthetic polymer formation, classification of polymerization processes, Relationships between functionality, extent of reaction and degree of polymerization. Bi-functional systems, Poly-functional systems.

3. Kinetics of Polymerization : *(8L)*
- Mechanism and kinetics of step growth, radical chain growth, ionic chain (both cationic and anionic) and coordination polymerizations, Mechanism and kinetics of copolymerization, polymerization techniques.

4. Crystallization and crystallinity : *(4L)*
- Determination of crystalline melting point and degree of crystallinity, Morphology of crystalline polymers, Factors affecting crystalline melting point.

5. Nature and structure of polymers : *(6L)*
- Structure Property relationships. (M_n, M_w etc) by end group analysis, viscometry, light scattering and osmotic pressure methods. Molecular weight distribution and its significance. Polydispersity index.

6. Glass transition temperature (T_g) and determinationof T_g : *(4L)*
- Free volume theory, WLF equation, Factors affecting glass transition temperature (T_g).

7. Polymer Solution : *(5L)*
- Criteria for polymer solubility, Solubility parameter, Thermodynamics of polymer solutions, entropy, enthalpy, and free energy change of mixing of polymers solutions, Flory-Huggins theory, Lower and Upper critical solution temperatures.

8. Properties of Polymer : *(18L)*

(Physical, Thermal, Flow & Mechanical Properties)

- Brief introduction to preparation, structure, properties and application of the following polymers: polyolefins, polystyrene and styrene copolymers, poly(vinyl chloride) and related polymers, poly(vinyl acetate) and related polymers, acrylic polymers, fluoro polymers. Polyamides and related polymers. Phenol formaldehyde resins (Bakelite, Novalac), polyurethanes, silicone polymers, polydienes. Polycarbonates, Conducting Polymers, [polyacetylene, polyaniline, poly(p-phenylene sulphide polypyrrole, poly-thiophene)].

9. Biopolymers : *(5L)*

- Polysaccharides, peptides, proteins and nucleic acids.

WEST BENGAL STATE UNIVERSITY

SEMESTER-VI

CEMADSE06T : POLYMER CHEMISTRY

(Credits : Theory-06, Practicals-02)

Theory : 60 Lectures

Marks : 50

Introduction and history of polymeric materials : *(04 Lectures) Marks : 04*

- Different schemes of classification of polymers, Polymer nomenclature, Molecular forces and chemical bonding in polymers, Texture of Polymers.

Functionality and its importance : *(08 Lectures) Marks : 06*

- Criteria for synthetic polymer formation, classification of polymerization processes, Relationships between func- tionality, extent of reaction and degree of polymerization. Bifunctional systems, Poly-functional systems.

Kinetics of Polymerization : *(08 Lectures) Marks : 06*

- Mechanism and kinetics of step growth, radical chain growth, ionic chain (both cationic and anionic) and coordination polymerizations, Mechanism and kinetics of copolymerization, polymerization techniques.

Crystallization and crystallinity : *(04 Lectures) Marks : 04*

- Determination of crystalline melting point and degree of crystallinity, Morphology of crystalline polymers, Factors affecting crystalline melting point.

Nature and structure of polymers : *(04 Lectures) Marks : 04*

- Structure Property relationships.

Determination of molecular weight of polymers : *(08 Lectures) Marks: 06*

- (M_n, M_w etc) by end group analysis, viscometry, light scattering and osmotic pressure methods. Molecular weight distribution and its significance. Polydispersity index.

Glass transition temperature (T_g) and determination of T_g : *(08 Lectures) Marks : 04*

- Free volume theory, WLF equation, Factors affecting glass transition temperature (T_g).

Polymer Solution : *(08 Lectures) Marks : 06*

- Criteria for polymer solubility, Solubility parameter, Thermodynamics of polymer solutions, entropy, enthalpy, and free energy change of mixing of polymers solutions, Flory-Huggins theory, Lower and Upper critical solution temperatures.

Properties of Polymer : *(10 Lectures) Marks : 10*

(Physical, thermal, Flow & Mechanical Properties).

- Brief introduction to preparation, structure, properties and application of the following polymers: polyolefins, polystyrene and styrene copolymers, poly(vinyl chloride) and related polymers, poly(vinyl acetate) and related polymers, acrylic polymers, fluoro polymers. polyamides and related polymers. Phenol formaldehyde resins (Bakelite, Novalac), polyurethanes, silicone polymers, polydienes. Polycarbonates, Conducting Polymers, [polyacetylene, polyaniline, poly(p-phenylene sulphide polypyrrole, polythiophene)].

VIDYASAGAR UNIVERSITY

B. Sc. (H)

SEMESTER-VI

DSE-4 : Polymer Chemistry

Credits 06

DSE4T : Polymer Chemistry

Credits 04

Introduction and history of polymeric materials :

- Different schemes of classification of polymers, Polymer nomenclature, Molecular forces and chemical bonding in polymers, Texture of Polymers.

Functionality and its importance :

- Criteria for synthetic polymer formation, classification of polymerization processes, Relationships between functionality, extent of reaction and degree of polymerization. Bi-functional systems, Poly-functional systems.

Kinetics of Polymerization :

- Mechanism and kinetics of step growth, radical chain growth, ionic chain (both cationic and anionic) and coordination polymerizations, Mechanism and kinetics of copolymerization, polymerization techniques.

Crystallization and crystallinity :

- Determination of crystalline melting point and degree of crystallinity, Morphology of crystalline polymers, Factors affecting crystalline melting point.

Nature and structure of polymers :

- Structure Property relationships.

Determination of molecular weight of polymers :

- (M_n, M_w, etc) by end group analysis, viscometry, light scattering and osmotic pressure methods. Molecular weight distribution and its significance. Polydispersity index.

Glass transition temperature (T_g) and determination of T_g :

- Free volume theory, WLF equation, Factors affecting glass transition temperature (T_g).

Polymer Solution :

- Criteria for polymer solubility, Solubility parameter, Thermodynamics of polymer solutions, entropy, enthalpy, and free energy change of mixing of polymers solutions, Flory-Huggins theory, Lower and Upper critical solution temperatures.

Properties of Polymer :

(Physical, thermal, Flow & Mechanical Properties).

- Brief introduction to preparation, structure, properties and application of the following polymers: polyolefins, polystyrene and styrene copolymers, poly(vinyl chloride) and related polymers, poly(vinyl acetate) and related polymers, acrylic polymers, fluoro polymers, polyamides and related polymers. Phenol formaldehyde resins (Bakelite, Novalac), polyurethanes, silicone polymers, polydienes. Polycarbonates, Conducting Polymers, [polyacetylene, polyaniline, poly(p-phenylene sulphide polypyrrole, poly-thiophene)].

SEMESTER-IV

CEM – 402

(Physical Special)

Unit-2 : Macromolecular & Biopolymers :

- Molecular weight of polymers, molecular weight determination by viscosity, osmometry, light scattering, diffusion and ultracentrifugation methods. Thermodynamics of polymer solutions. Kinetics of polymerization.

UNIVERSITY OF NORTH BENGAL

SEMESTER-VI

CHEMISTRY-DSE : POLYMER CHEMISTRY

(Credits : Theory-06, Practicals-02)

Theory : 60 Lectures

Introduction and history of polymeric materials : *(2 Lectures)*

- Different schemes of classification of polymers, Polymer nomenclature, Molecular forces and chemical bonding in polymers, Texture of Polymers. Classifications including di-, tri-, and amphiphilic polymers.

Functionality and its importance : *(10 Lectures)*

- Addition and Condensation –Mechanism of Cationic,aninonic and free radical addition polomerization.
- Criteria for synthetic polymer formation, classification of polymerization processes, Relationships between functionality, extent of reaction and degree of poly- merization. Bi-functional systems, Polyfunctional systems.

Kinetics of Polymerization : *(6 Lectures)*

- Mechanism and kinetics of step growth, radical chain growth, ionic chain (both cationic and anionic) and coordination polymerizations, Mechanism and kinetics of copolymerization, polymerization techniques. Metallocene-based Ziegler-Natta polymerisation of alkenes; Preparation and applications of plastics – thermosetting (phenol-formaldehyde, Polyurethanes) and thermosoftening (PVC, polythene).

Crystallization and crystallinity : *(4 Lectures)*

- Determination of crystalline melting point and degree of crystallinity, Morphology of crystalline polymers, Factors affecting crystalline melting point.

Nature and structure of polymers :
(2 Lectures)

- Structure Property relationships.

Determination of molecular weight of polymers : *(8 Lectures)*

- (M_n, M_w, etc) by end group analysis, viscometry, light scattering and osmotic pressure methods. Molecular weight distribution and its significance.
- Polydispersity index.

Glass transition temperature (T_g) and determination of T_g : *(6 Lectures)*

- Free volume theory, WLF equation, Factors affecting glass transition temperature (T_g).

Polymer Solution : *(6 Lectures)*

- Criteria for polymer solubility, Solubility parameter, Thermodynamics of polymer solutions, entropy, enthalpy, and free energy change of mixing of polymers solutions, Flory-Huggins theory, Lower and Upper critical solution temperatures.

Properties of Polymers *(10 Lectures)*
(Physical, thermal, Flow & Mechanical Properties)

- Brief introduction to preparation, structure, properties and application of the following polymers : polyolefins, polystyrene and styrene copolymers, poly(vinyl chloride) and related polymers, poly(vinyl acetate) and related polymers, acrylic polymers, fluoro polymers. polyamides and related polymers, Phenol formaldehyde resins (Bakelite, Novalac), polyurethanes, silicone polymers, polydienes. Polycarbonates, Conducting Polymers, [polyacetylene, polyaniline, poly(p-phenylene sulphide polypyrrole, poly-thiophene)].

Fabrics : *(6 Lectures)*

- Natural and synthetic (acrylic, polyamido, polyester); Rubbers – natural and synthetic : Buna-S, Chloroprene and Neoprene; Vulcanization; Polymer additives; Introduction to liquid crystal polymers; Biodegradable and conducting polymers with examples.

UNIVERSITY OF KALYANI

SEMESTER-V

CHEMHTDSE-1A

Theory : Introduction, Functionality and its importance, Kinetics of Polymerization, Crystallization and crystallinity, Nature and structure of polymers, molecular weight of polymers, T_g, Solubility and Properties. *4 Credits*

1. **Introduction and history of polymeric materials :** *(4L)*

- Different schemes of classification of polymers, Polymer nomenclature, Molecular forces and chemical bonding in polymers, Texture of Polymers.

2. **Functionality and its importance :** *(6L)*

- Criteria for synthetic polymer formation, classification of polymerization processes, elationships between functionality, extent of reaction and degree of poly- merization. Bi-functional systems, Poly-functional systems.

3. **Kinetics of Polymerization :** *(8L)*

- Mechanism and kinetics of step growth, radical chain growth, ionic chain (both cationic and anionic) and coordination polymerizations.

4. **Crystallization and crystallinity :** *(4L)*

- Determination of crystalline melting point and degree of crystallinity, Morphology of crystalline polymers, Factors affecting crystalline melting point.

5. **Nature and structure of polymers :** *(4L)*

- Structure Property relationships.

6. **Determination of molecular weight of polymers :** *(6L)*

- (M_n, M_w, etc) by end group analysis, viscometry, light scattering and osmotic pressure methods. Molecular weight distribution and its significance.
- Polydispersity index.

7. **Glass transition temperature (T_g) and determination of T_g :** *(4L)*

- Free volume theory, WLF equation, Factors affecting glass transition temperature (T_g).

8. **Polymer Solution :** *(10L)*

- Criteria for polymer solubility, Solubility parameter, Thermodynamics of polymer solutions, entropy, enthalpy, and free energy change of mixing of polymers solutions, Lower and Upper critical solution temperatures.

9. **Properties of Polymer :** *(14L)*
(Physical, thermal, Flow & Mechanical Properties)

- Brief introduction to preparation, structure, properties and application of the following polymers : polyolefins, polystyrene and styrene copolymers, poly(vinyl chloride) and related polymers, poly(vinyl acetate) and related polymers, acrylic polymers, fluoro polymers. Polyamides and related polymers. Phenol formaldehyde resins (Bakelite, Novalac), polyurethanes, silicone polymers, polydienes. Polycarbonates, Conducting Polymers, [polyacetylene, polyaniline, poly(p-phenylene sulphide polypyrrole, polythiophene)].

SEMESTER-VI

Course Code : DSE-3

Course Title : Polymer Chemistry (Theo)
4 Credits

Introduction and history of polymeric materials :
- Different schemes of classification of polymers, Polymer nomenclature, Molecular forces and chemical bonding in polymers, Texture of Polymers.

Functionality and its importance :
- Criteria for synthetic polymer formation, classification of polymerization processes, Relationships between functionality, extent of reaction and degree of polymerization. Bifunctional systems, Poly-functional systems.

Kinetics of Polymerization :
- Mechanism and kinetics of step growth, radical chain growth, kinetics of copolymerization, polymerization techniques.

Determination of molecular weight of polymers :
- M_n, M_w, etc. (by end group analysis), viscometry, osmotic pressure methods. Molecular weight distribution and its significance. Polydispersity index.

Glass transition temperature (T_g) and determination of T_g :
- Free volume theory, WLF equation, Factors affecting glass transition temperature (T_g).

Polymer Solution :
- Criteria for polymer solubility, Solubility parameter, Thermodynamics of polymer solutions, entropy, enthalpy, and free energy change of mixing of polymers solutions.

Properties of Polymer :
(Physical, thermal, Flow & Mechanical Properties)
- Brief introduction to preparation, structure, properties and application of the following polymers : polyolefins, polystyrene and styrene copolymers, poly(vinyl chloride) and related polymers. Polyamides and related polymers. Phenol formaldehyde resins (Bakelite, Novalac). Polycarbonates, Conducting Polymers, [polyacetylene, polyaniline, poly(p-phenylene sulphide polypyrrole, polythiophene)].

SEMESTER-VI

CHEMISTRY-DSE : POLYMER CHEMISTRY
(Credits : Theory-06, Practicals-02)
Theory : 60 Lectures

Introduction and history of polymeric materials :
- Different schemes of classification of polymers, Polymer nomenclature, Molecular forces and chemical bonding in polymers, Texture of Polymers.

Functionality and its importance :
- Criteria for synthetic polymer formation, classification of polymerization processes, Relationships between functionality, extent of reaction and degree of polymerization. Bifunctional systems, Poly-functional systems.

Kinetics of Polymerization :
- Mechanism and kinetics of step growth, radical chain growth, ionic chain (both cationic and anionic) and coordination polymerizations, Mechanism and kinetics of copolymerization, polymerization techniques.

Crystallization and crystallinity :
- Determination of crystalline melting point and degree of crystallinity, Morphology of crystalline polymers, Factors affecting crystalline melting point.

Nature and structure of polymers :
- Structure Property relationships.

Determination of molecular weight of polymers :
- (M_n, M_w, etc) by end group analysis, viscometry, light scattering and osmotic pressure methods. Molecular weight distribution and its significance. Polydispersity index.

Glass transition temperature (T_g) and determination of T_g :
- Free volume theory, WLF equation, Factors affecting glass transition temperature (T_g).

Polymer Solution :
- Criteria for polymer solubility, Solubility parameter, Thermodynamics of polymer solutions, entropy, enthalpy, and free energy change of mixing of polymers solutions, Flory-Huggins theory, Lower and Upper critical solution temperatures.

Properties of Polymers :
(Physical, thermal, Flow & Mechanical Properties)

- Brief introduction to preparation, structure, properties and application of the following polymers: polyolefins, polystyrene and styrene copolymers, poly(vinyl chloride) and related polymers, poly(vinyl acetate) and related polymers, acrylic polymers, fluoro polymers, polyamides and related polymers. Phenol formaldehyde resins (Bakelite, Novalac), polyurethanes, silicone polymers, polydienes, Polycarbonates, Conducting Polymers, [polyacetylene, polyaniline, poly(p-phenylene sulphide polypyrrole, polythiophene)].

SIDHO-KANHO-BIRSHA UNIVERSITY

SEMESTER-VI

Title : Polymer Chemistry
Theory

Introduction and history of polymeric materials : *(4L)*

- Different schemes of classification of polymers, Polymer nomenclature, Molecular forces and chemical bonding in polymers, Texture of Polymers.

Functionality and its importance : *(4L)*

- Criteria for synthetic polymer formation, classification of polymerization processes, Relationships between functionality, extent of reaction and degree of polymerization. Bi-functional systems, Poly-functional systems.

Kinetics of Polymerization : *(6L)*

- Mechanism and kinetics of step growth, radical chain growth, ionic chain (both cationic and anionic) and coordination polymerizations, Mechanism and kinetics of copolymerization, polymerization techniques.

Crystallization and crystallinity : *(6L)*

- Determination of crystalline melting point and degree of crystallinity, Morphology of crystalline polymers, Factors affecting crystalline melting point.

Nature and structure of polymers : *(3L)*

- Structure Property relationships.

Determination of molecular weight of polymers : *(7L)*

- (M_n, M_w, etc) by end group analysis, viscometry, light scattering and osmotic pressure methods. Molecular weight distribution and its significance. Polydispersity index.

Glass transition temperature (T_g) and determination of T_g : *(5L)*

- Free volume theory, WLF equation, Factors affecting glass transition temperature (T_g).

Polymer Solution : *(10 L)*

- Criteria for polymer solubility, Solubility parameter, Thermodynamics of polymer solutions, entropy, enthalpy, and free energy change of mixing of polymers solutions, Flory–Huggins theory, Lower and Upper critical solution temperatures.

Properties of Polymer : *(15L)*
(Physical, thermal, Flow & Mechanical Properties)

- Brief introduction to preparation, structure, properties and application of the following polymers : polyolefins, polystyrene and styrene copolymers, poly(vinyl chloride) and related polymers, poly(vinyl acetate) and related polymers, acrylic polymers, fluoro polymers, Polyamides and related polymers. Phenol formaldehyde resins (Bakelite, Novalac), polyurethanes, silicone polymers, polydienes, Polycarbonates, Conducting Polymers, [polyacetylene, polyaniline, poly(p-phenylene sulphide polypyrrole, polythiophene)].

BANKURA UNIVERSITY

SEMESTER-VI

3.7 DSE T4 – Polymer Chemistry
4 Credits

Introduction and history of polymeric materials :

- Different schemes of classification of polymers, Polymer nomenclature, Molecular forces and chemical bondingin polymers, Texture of Polymers.

Functionality and its importance :

- Criteria for synthetic polymer formation, classification of polymerization processes, Relationships between functionality, extent of reaction and degree of polymerization. Bifunctional systems, Poly-functional systems.

Kinetics of Polymerization :

- Mechanism and kinetics of step growth, radical chain growth, ionic chain (both cationic and anionic) and coordination polymerizations, Mechanism and kinetics of copolymerization, polymerization techniques.

Crystallization and crystallinity :

- Determination of crystalline melting point and degree of crystallinity, Morphology of crystalline polymers, Factors affecting crystalline melting point.

Nature and structure of polymers

- Structure Property relationships.

Determination of molecular weight of polymers :

- (M_n, M_w, etc) by end group analysis, viscometry, light scattering and osmotic pressure methods. Molecular weight distribution and its significance. Polydispersity index.

Glass transition temperature (T_g) and determination of T_g

- Free volume theory, WLF equation, Factors affecting glass transition temperature (T_g).

Polymer Solution :

- Criteria for polymer solubility, Solubility parameter, Thermodynamics of polymer solutions, entropy, enthalpy, and free energy change of mixing of polymers solutions, Flory-Huggins theory, Lower and Upper critical solution temperatures.

Properties of Polymer :

(Physical, thermal, Flow & Mechanical Properties)

- Brief introduction to preparation, structure, properties and application of the following polymers : polyolefins, polystyrene and styrene copolymers, poly(vinyl chloride) and related polymers, poly(vinyl acetate) and related polymers, acrylic polymers, fluoro polymers, Polyamides and related polymers. Phenol formaldehyde resins (Bakelite, Novalac), poly-urethanes, silicone polymers, polydienes, Poly-carbonates, Conducting Polymers, [polyacetylene, polyaniline, poly(p-phenylene sulphide polypyrrole, polythiophene)].

UNIVERSITY OF GOUR BANGA

SEMESTER-V

CEMHTDSE- 2B

Theory : Introduction, Functionality and its importance, Kinetics of Polymerization, Crystallization and crystallinity, Nature and structure of polymers, molecular weight of polymers, glass transition temperature (T_g) , polymer solution, Properties of polymer. *4 Credits*

1. Introduction and history of polymeric materials : *(4L)*

- Different schemes of classification of polymers, Polymer nomenclature, Molecular forces and chemical bonding in polymers, Texture of Polymers.

2. Functionality and its importance : *(6L)*

- Criteria for synthetic polymer formation, classification of polymerization processes, elationships between functionality, extent of reaction and degree of polymerization. Bi-functional systems, Poly- functional systems.

3. Kinetics of Polymerization : *(8L)*

- Mechanism and kinetics of step growth, radical chain growth, ionic chain (both cationic and anionic) and coordination polymerizations.

4. Crystallization and crystallinity : *(4L)*

- Determination of crystalline melting point and degree of crystallinity, Morphology of crystalline polymers, Factors affecting crystalline melting point.

5. Nature and structure of polymers : *(4L)*

- Structure Property relationships.

6. Determination of molecular weight of polymers : *(6L)*

- (M_n, M_w, etc) by end group analysis, viscometry, light scattering and osmotic pressure methods. Molecular weight distribution and its significance. Polydispersity index.

7. Glass transition temperature (T_g) and determination of T_g : *(4L)*

- Free volume theory, WLF equation, Factors affecting glass transition temperature (T_g).

8. Polymer Solution : *(10L)*

- Criteria for polymer solubility, Solubility parameter, Thermodynamics of polymer solutions, entropy, enthalpy, and free energy change of mixing of polymers solutions, Flory-Huggins theory, Lower and Upper critical solution temperatures.

9. Properties of Polymer : *(14L)*

(Physical, thermal, Flow & Mechanical Properties)

- Brief introduction to preparation, structure, properties and application of the following polymers :
Polyolefins, polystyrene and styrene copolymers, poly(vinyl chloride) and related polymers, poly(vinyl acetate) and related polymers, acrylic polymers, fluoro polymers, Polyamides and related polymers. Phenol formaldehyde resins (Bakelite, Novalac), poly-urethanes, silicone polymers, polydienes, Poly-carbonates, Conducting Polymers, [polyacetylene, polyaniline, poly(p-phenylene sulphide polypyrrole, polythiophene)].

M. Sc.

SEMESTER-III

MSCCHEMC303 : Advanced Physical Chemistry General

Theoretical Papers

(For Each, Full Marks : 50; Credit : 4)

UNIT-II

3. Chemistry of Polymers : *(08 lectures)*

● Basic Concepts, classification, nomenclature, molecular weights, molecular weight distribution Methods for determination of Molecular weights, viscosity molecular weight, intrinsic viscosity, Mark-Houwink relationships, Glass transition temperature, Polymerization reaction, kinetics of free radical and condensation polymer. Graft polymerization. Morphology and crystallinity of polymer by TGA and SEM analysis. Criteria for polymer solubility.Thermodynamics of polymer solutions. Theta temperature. Flory-Huggins model, dilute polymer solution. Excluded volume.

Course objectives :

* To understand theory of polymerization and determination of molecular weight of polymer.
* To understand kinetics of polymerization.
* To understand theories of polymer solutions.

Learning outcomes :

After completing the course the student will be expected to be able to :

* determine molecular weight of polymer.
* determine rate / degree / extent of polymerization.
* explain behavior of polymer in solution.

VISVA-BHARATI UNIVERSITY

M. Sc.

SEMESTER-I

Theoretical

CH 702 : Inorganic Chemistry (Core)

Full Marks : 50 (40 + 10); Credit point : 4

2. Polymer Chemistry-1 : *(12L)*

● Basics : Importance of polymers, Basic concepts of monomers, repeat units, degree of polymerization, linear, branched and network polymers, classification of polymers, polymerization-condensation, addition, radical chain-ionic and coordination, copolymerization; Polymer characterization- number, weight and viscosity average molecular weights, polydispersity index and molecular weight distribution, measurement of molecular weight by viscosity method; Structure and property- the glass transition temperature, relationship between T_g and T_m, factors controlling T_g; Functional polymers-fire retarding polymers and electrically conducting polymers; Biomedical polymers- contact lens, dental polymers, artificial heart, kidney, skin and blood cells.

SEMESTER-IV

Theoretical

CH 1019 : Elective-3 (Physical)

Full Marks : 50 (40 + 10)

Credit point : 4

3. Physcial Chemistry of Polymers : *(18L)*

● Polymerization reaction, kinetics of free radical and condensation polymer. Graft polymerization. Morphology and crystallinity of polymer by TGA and SEM analysis. Molecular weight determination of polymer by light scattering method and GPC method.

● Criteria for polymer solubility. Thermodynamics of polymer solutions. Good and bad solvents. Theta temperature. Flory-Huggins model, dilute polymer solution. Excluded volume.

DIAMOND HARBOUR WOMENS' UNIVERSITY

M. Sc.

SEMESTER-III

Course ID : Chem/ThSP/303

Unit-3 : Polymer chemistry

● Classification of polymers; kinetics of polymerization. Mean molar masses of polymers and the various methods of determinations; nature of distributions about the mean. Thermodynamics of polymer solution : Polymer conformation.

CONTENTS

Chapter 9 — Preparation, Structure, Properties and Application of Polymers 116-158

History and Classification of Polymers

1.1 History of Polymeric Materials

Natural rubber is the polymer produced naturally from the native Brazilian plant *Hevea brasiliensis*. The first scientific report on rubber in the form of paper was published in 1755 by **Charles Marie de La Condamine**, a French explorer. In 1770 **Joseph Priestley**, an English chemist observed rubber was excellently adapted to the purpose of wiping the mark of black lead pencil from paper. John Gough, the English natural and experimental philosopher, described thermoelastic properties of rubber using his three experiments in 1804. After that in 1826 English chemist Michael Faraday determined the elemental composition of natural rubber with the empirical formula C_5H_8, along with $\leq 4\%$ protein and $\leq 4\%$ acetone-soluble materials. Johann Eduard Simon in 1839 isolated styrol (now it is styrene) by distillation storax resin with sodium carbonate. By the way, it was the first recorded occurrence of polymerization. After a few years later, Blyth and Hoffman reported first photopolymerization of metastyrol in 1845. After that Marcellin Berthelot reported the thermal and catalytic polymerization of pinene in 1853. Bakelite was the first synthetic plastic invented by Leo Baekeland in 1907. In 1920 Hermann Staudinger gave the concept of macromolecules as large molecules linked together by covalent bonds. Wallace Carothers in this equation relates the degree of polymerization to the conversion of monomer into polymer in 1930. Paul John Flory was a leading researcher in understanding the behaviour of polymers in solution.

The following scientists received the Nobel Prize in the field of polymer science.

Name of the Scientist/s	Nobel Prize (Year)
Hermann Staudinger	1953
Giulio Natta and Karl Ziegler	1963
Paul John Flory	1974

Name of the Scientist/s	Nobel Prize (Year)
Pierre-Gilles de Gennes	1991
Alan J. Heeger, Alan MacDiarmid, and Hideki Shirakawa	2000
John Bennett Fenn, Koichi Tanaka, and Kurt Wüthrich	2002
Robert Grubbs, Richard Schrock and Yves Chauvin	2005

1.2 Classification of Polymers

Polymers can be obtained from different sources and have different synthetic procedures, properties like structure, thermal responses, crystallinity, hydrophobicity, polarity, tacticity etc. and applicability. On the basis of above mention fact polymers have been classified into the following category

Nature of Classification	Polymers category
1. Sources of polymer	Natural, semi synthetic, synthetic organic, synthetic inorganic.
2. Types of Polymerization	Addition, condensation.
3. Structures of polymer	Linear, branched, cross-linked or network.
4. Molecular Forces	Elastomers, fibres, thermoplastic, thermosetting.
5. Crystallinity	Crystalline, amorphous.
6. Monomer	Homopolymers, heteropolymers.
7. Tacticity	Isotactic, syndiotactic, atactic.
8. Hydrophobicity	Hydrophobic, hydrophilic.

1.3 Definitions and Examples of Different Types of Polymer

- **Natural polymer :** Natural polymers are the bio-macromolecules which occur naturally in the environment. Generally natural polymers remain in the environment as Polynucleotides (DNA, RNA), Polysaccharides (Cellulose) and Proteins (hair, skin). These polymers are good for biomedical application. **Example :** Natural rubber, natural silk, cellulose, starch, proteins, Leather, RNA, DNA.

- **Semi synthetic polymer :** Synthetic polymers are those polymers which are obtained by synthesis of natural polymers or monomers. These are man made polymers. When the mother compound is a polymer & gives another polymer by simple modification (acetylation, nitration etc) then the derivatives of mother compound are called semi synthetic polymers. **Example :** Acetate rayon, cup ammonium silk, viscous rayon.

- **Synthetic organic polymer :** When the polymers are prepared by reacting with monomers, such polymers are called synthetic polymers. If the polymer backbone has carbon atom (must present) and other atoms (N, S, O etc) (may or may not present) then the polymer is called synthetic organic polymer. **Example :** Polyvinyl alcohol, polyethylene, polystyrene, polysulfone, PVC, bakelite, teflon.

- **Synthetic inorganic polymer :** When the synthetic polymers backbone contains phosphorus/silicon atom instead of carbon atom are called synthetic inorganic polymers. **Example :** Polysiloxanes, polyphosphazenes, and polysilanes.

- **Addition polymer :** This type of polymers are prepared by adding one monomer to the polymer chain by chain reaction. **Example :** Polyethylene, polypropylene, polystyrene.

- **Condensation polymer :** This type of polymers are prepared by condensation reactions between two different monomers (monomer should be bifunctional or polyfunctional) **Example :** Cellulose, poly(β-hydroxybutyric acid), polyester, Polyamides.

- **Linear polymer :** When the polymers have linear backbone in a single straight line are called linear polymers.

Pic 1.1 Liner Polymers

  Example : Polyethene, PVC, Nylons, polyesters.

- **Branched polymer :** When the polymers have a main chain along with one or more side chains, the polymers are called branched polymers. The length of main chain and side chains may vary.

Pic 1.2 Branched Polymers

  The length is also vary from one side chain to another side chain.

  Example : Polypropylene, amylopectin and glycogen.

- **Cross-linked or network polymer :** When the polymers chains are cross-linked each other, form a complex three dimensional network architecture called cross-linked or network polymers.

  Example : Bakelite, melamine, formaldehyde resins, vulcanised rubber.

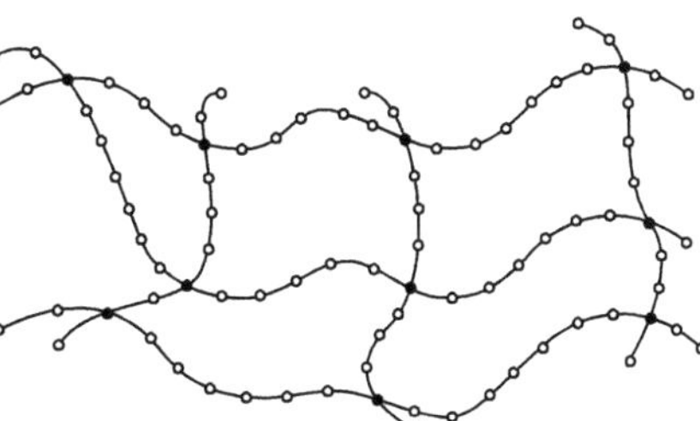

Pic 1.3 Cross-linked or network Polymers

- **Elastomeric polymer : See page 6 Example :** Vulcanized Rubber.

- **Fibrillar polymer :** See page 7 Example : Nylon-66.
- **Thermoplastic polymer :** See page 7 Example : Polyethylene, PVC, nylon.
- **Thermosetting polymer :** See page 7 Example : Bakelite.
- **Crystalline polymer :** In one word it the XRD spectra of polymers show sharp peak/s, the polymers are called crystalline polymers.
  Example : Polypropylene, Syndiotactic polystyrene, Nylon.
- **Amorphous polymer:** If the XRD-spectra of polymers show broad peak, the polymers are called amorphous polymers.
  Example : Poly(methyl methacrylate), Atactic polystyrene, Polycarbonate, Polyisoprene.
- **Homopolymers :** When the polymers are formed by single monomer units, called homopolymers. **Example :** Polystyrene, polyethylene.
- **Heteropolymers :** When the polymers are formed by two or more monomer units, called heteropolymers. **Example :** DNA, protein.
- **Isotactic polymer :** See page 5 Example : Polypropylene.
- **Syndiotactic polymer :** See page 5 Example : Syndiotactic polystyrene
- **Atactic polymer :** See page 5 Example : Atactic polystyrene
- **Hydrophobic polymer :** When a water droplet made a contact angle more than 90° on polymer surface is called hydrophobic polymers.
  Example: Polyethylene, polystyrene, polyvinylchloride, polytetrafluorethylene, polydimethylsiloxane.
- **Hydrophilic polymer :** When the water droplet contact angle is less than 90° on the polymers surface, the polymers are called hydrophilic polymers.
  Example : Polyvinyl alcohol, cellulose, polyamides, polyethylene glycol ethers, polyacrylic amides.

1.4 Polymer Nomenclature

Polymers are formed by one or more than one type of repeating unit. Let A and B are two different repeating monomer units then following types of polymers will be formed.

1. **Homopolymers :** $-(A)_n-$ or $-(B)_n-$
2. **Copolymers :**

i. **Alternating copolymer :** $-(A-B)_n-$ or $-(B-A)_n-$

ii. **Random copolymer :** $-(A-B-B-A-A-B-A-B-B-B)_n-$

iii. **Block copolymer :** $-(B-B-B-A-A-A-B-B-B-A-A-A)_n-$

iv. **Graft copolymer :** $-(A-A-A-A-A-A-A-A-A-A)_n-$
$$|||$$
$$(B)_n(B)_n(B)_n$$
$$|||$$

If the monomer unit (having two different functional groups) polymerized chiral centre will form. With the help of stereochemistry the structures of the polymer can be easily explained. Suppose the monomer units have R and H groups attached with each carbon (backbone) atom. In the final polymer when all the R group remain in one plane (above the plane or below the plane) is called **isotactic polymer**, when the R groups remain above the plane and below the plane or vice-verse consecutively then the polymer is called **syndiotactic polymer**. But when the R groups remain haphazardly above the plane or below the plane in the polymer chain, **atactic polymer** is formed. Thus according to tacticity polymers can be following.

Isotactic polymers: (all R groups are above the plane)

Syndiotactic polymers: (R groups are above the plane and below the plane alternatively)

Atactic polymers: (R groups are haphazardly above the plane and below the plane)

1.5 Follow the Simple Procedures for Polymers Nomenclature

Homopolymers

(A) For polymers made from simple single repeating unit monomer having no structural features use the prefix poly followed by monomer name
Examples: polyaniline, polypyrrole, polythiophene.

(B) For polymers made from single repeating unit monomer having structural features use the prefix poly followed by monomer name in bracket
Examples: poly(4-chlorostyrene), poly(N-isopropylacrylamide), poly(methyl methacrylate).

Copolymers

(A) For unspecified copolymers made from two different repeating unit monomers having no specific arrangement use the prefix poly followed by two monomers name separated by the term '*-co-*' in bracket

Examples : poly(styrene-*co*-butadiene), poly(styrene-*co*-isoprene).

(B) For alternating copolymers made from two different repeating unit monomers having alternate arrangement use the prefix poly followed by two monomers name separated by the term '-*alt*-' in bracket

Examples: poly(butadiene-*alt*-acrylonitrile), poly [styrene-*alt*-(maleic anhydride)].

(C) For random copolymers made from two different repeating unit monomers having random arrangement use the prefix poly followed by two monomers name separated by the term '-*ran*-' in bracket

Examples: poly[(methyl methacrylate)-*ran*-(butyl acrylate)].

(D) For block copolymers made from two different repeating unit monomers having block arrangement use the prefix poly for both monomers name separated by the term '-*block*-'

Examples: poly(buta-1,3-diene)-*block*-poly(ethene-co-propene).

(E) For graft copolymers made from two different repeating unit monomers having graft arrangement use the prefix poly for both monomers name separated by the term '-*graft*-'.

Examples: polystyrene-*graft*-poly(ethylene oxide).

Beside that there are non-linear copolymers and polymer assemblies which have different shape and structural arrangement with the name blend copolymers, comb copolymers, branch copolymers, network copolymers, star copolymers, cyclic copolymers, complex copolymers and others. The terms used for the nomenclature of these polymers are -*blend*-, -*comb*-, *branch*-, *net*-, *star*-, *cyclo*- and −*compl*- respectively.

1.6 Molecular Forces and Chemical Bonding in Polymers

Polymers exhibit the same intermolecular forces and chemical bonding as like small molecules. van der Waals, forces, hydrogen bonding and dipole-dipole interactions are the three main forces acting on polymers chain. Beside that in ionic polymers ionic interactions are involved. On the basis of inter-molecular forces between monomer molecules, the polymers have been classified into four types namely *Elastomers*, *fibres*, *thermoplastic* and *thermosetting*.

Elastomer

The polymer chains are held together by the weakest intermolecular forces like van der Waals, forces. When an elongated or shear force is applied to an elastomer, it stretches or deforms but when the force is removed it reverts back to its original shape and size.

Example : natural rubber, polyisoprene, polybutadiene, nitrile rubbers, ethylene propylene rubber, polyester.

 ## Fibres

The polymer chains are held together by strong intermolecular hydrogen bonding and/or dipole-dipole interactions. As polymer chains are aligned in fibres so it will not stretch like an elastomer. Thus fibres have high tensile strength, lower elasticity and sharp melting point.

Example : Silk, nylon, polyester, phenol-formaldehyde, PVC fibre, polyamide, kevlar, Nomex.

 ## Thermoplastic Polymers

The polymer chains are held together by weak van der Waals, forces. The polymer chains are linear with no or very few cross linkages. Polymers have the characteristic of softening on heating and can be converted into desired shape because of the flexible nature at high temperature.

Example : Polythene, PVC, polystyrene, polypropylene, polycarbonate, poly ether, sulfone, polyetherimide, teflon, poly oxymethylene.

 ## Thermosetting polymers

The polymer chains cross-link and/or undergo other chemical change on heating, resulting in infusible and insoluble hard mass. For the preparation of these types of polymers only one attempt will be given no second chance.

Example : Bakelite, Epoxy resins, polycarbonates, cyanate ester, melamine, polyester resin, polyurethane, vulcanized rubber.

1.7 Texture of Polymers

It can be divided into macroscopically (using naked eye) and microscopically (using electron microscope or optical microscope) polymers texture. Macroscopically polymers texture are powdery or crystal like and microscopically are aggregated or fibrillar. Beside that using different scientific techniques (preparation and processing) polymers can be processed into desired texture. Examples of macroscopic textures are sheet, pipe, disc, plate etc. and for microscopic textures (morphology) are microsphere, hollow microsphere, nanotube, hollow nanotube, nanoparticle, microparticle and so on.

Pic 1.4 Powder

Pic 1.5 Crystal

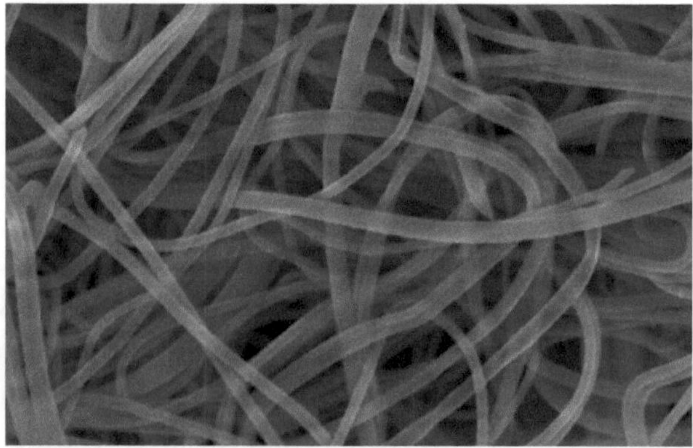

Pic 1.6 Nanofibre

QUESTION & ANSWER

1. Write the Nobel laureates in polymer chemistry.

Ans. Polymer is a vast subject and many researchers working in different fields. The researchers who got the Nobel Prize are as follows. Hermann Staudinger got the Nobel Prize in 1953 for the demonstration of the existence of macromolecules, which he characterized as polymers.After 10 years later in 1963 Karl Ziegler and Giulio Natta jointly awarded the Nobel Prize for their discoveries in the field of the chemistry and technology of high polymers. Paul John Flory received the Nobel Prize in 1974 for his fundamental achievements, both theoretical and experimental, in the physical chemistry of the macromolecules.Pierre-Gilles de Gennes was awarded the Nobel Prize for discovering the methods developed for studying order phenomena in simple systems that can be generalized to more complex forms of matter, in particular to liquid crystals and polymers in 1991. In 2000, the Nobel Prize was awarded jointly to Alan J. Heeger, Alan G. MacDiarmid and Hideki Shirakawa for the discovery and development of conductive polymers. The Nobel Prize in 2002 was awarded for the development of methods for identification and structure analysis of biological macromolecules with one half jointly to John B. Fenn and Koichi Tanaka for their development of soft desorption ionisation methods for mass spectrometric analysis of biological macromolecules and the other half to Kurt Wüthrich for his development of nuclear magnetic resonance spectroscopy for determining the three-dimensional structure of biological macromolecules in solution. In 2005,the Nobel Prize was awarded jointly to Yves Chauvin, Robert H. Grubbs and Richard R. Schrock for the development of the metathesis method in organic synthesis.

2. Classify polymer with examples.

Ans. Polymers have been classified on the basis of sources of polymer, types of polymerization process by which polymer formed, structures of polymer formed, molecular forces acting on polymer, crystallinity of the prepared polymer, number of monomer present in the polymer, tacticity of the prepared polymer and hydrophobicity of the prepared polymer.

▶ On the basis of sources, polymer have been categorized into natural, semi synthetic, synthetic organic and synthetic inorganic polymer. The examples are as follows

Natural polymer: Natural rubber, natural silk, cellulose, starch, proteins, Leather, RNA, DNA.

Semi synthetic polymer: Acetate rayon, cup ammonium silk, viscous rayon.

Synthetic organic polymer: Polyvinyl alcohol, polyethylene, polystyrene, polysulfone, PVC, bakelite, teflon.

Synthetic inorganic polymer: Polysiloxanes, polyphosphazenes, and polysilanes.

▶ On the basis of the polymerization process, polymer have been categorized into addition and condensation polymer. The examples are as follows

Addition polymer: Polyethylene, polypropylene, polystyrene.

Condensation polymer: Cellulose, poly(β-hydroxybutyric acid), polyester, Polyamides.

▶ On the basis of structures of polymer, polymer have been categorized into linear, branched and cross-linked or network polymer. The examples are as follows

Linear polymer: Polyethene, PVC, Nylons, polyesters.

Branched polymer: Polypropylene, amylopectin and glycogen.

Cross-linked or network polymer: Bakelite, melamine, formaldehyde resins, vulcanised rubber.

▶ On the basis of molecular forces, polymer have been categorized into elastomers, fibres, thermoplastic and thermosetting polymer. The examples are as follows

Elastomeric polymer: Vulcanized Rubber.

Fibrillar polymer: Nylon-66.

Thermoplastic polymer: Polyethylene, PVC, nylon.

Thermosetting polymer: Bakelite.

▶ On the basis of crystallinity, polymer have been categorized into crystalline and amorphous polymer. The examples are as follows

Crystalline polymer: Polypropylene, Syndiotactic polystyrene, Nylon.

Amorphous polymer: Poly(methyl methacrylate), Atactic polystyrene, Polycarbonate, Polyisoprene.

▶ On the basis of monomer, polymer have been categorized into homopolymers and heteropolymers. The example are as follows

Homopolymers: Polystyrene, polyethylene.

Heteropolymers: DNA, protein.

▶ On the basis of tacticity, polymer have been categorized into isotactic, syndiotactic and atactic polymers. The examples are as follows

Isotactic polymer: Isotactic Polypropylene.

Syndiotactic polymer: Syndiotactic polystyrene

Atactic polymer: Atactic polystyrene

▶ On the basis of hydrophobicity, polymer have been categorized into hydrophobic, hydrophilic polymers. The examples are as follows

Hydrophobic polymer: Polyethylene, polystyrene, polyvinylchloride, polytetrafluorethylene, polydimethylsiloxane.

Hydrophilic polymer: Polyvinyl alcohol, cellulose, polyamides, polyethylene glycol ethers, polyacrylic amides.

3. Draw the structure of polystyrene on the basis of tacticity.

Ans. According to the tacticity, polystyrene forms three different structures namely isotactic polystyrene, syndiotactic polystyrene and atactic polystyrene.

Isotactic polystyrene :

Syndiotactic polystyrene :

Atactic polystyrene :

4. What types of molecular forces are present in polymers?

Ans. Polymers exhibit the same molecular forces as small molecules. Van der Waals forces, hydrogen bonding and dipole-dipole interactions are the three main forces acting on polymer chain. Beside that in ionic polymers ionic interactions are involved. On the basis of inter-molecular forces between monomer molecules, the polymers have been classified into four types namely Elastomers, fibres, thermoplastic and thermosetting.

5. Give examples of (a) commodity plastics, (b) engineering plastics, (c) elastomers, (d) transparent plastics, (e) thermoset resins for composites.

Ans. (a) Commodity plastics: Polyethylene (PE), polyvinyl chloride (PVC) and polypropylene (PP).

(b) Engineering plastics: Acrylonitrile butadiene styrene (ABS), polycarbonates and polyamides (nylons).

(c) Elastomers: Natural rubbers, styrene-butadiene block copolymers, polyisoprene, polybutadiene, ethylene propylene rubber, ethylene propylene diene rubber, silicone elastomers, fluoro-elastomers, polyurethane elastomers, and nitrile rubbers.

(d) Transparent plastics: Polyethylene terephthalate (PET), polyvinyl chloride (PVC), polypropylene (PP), polystyrene (PS), polycarbonate (PC), polymethyl methacrylate (PMMA) and others (polyamide, ABS & SAN, Polyethylene, TPU).

(e) Thermoset resins: Polyester, epoxy, phenolic, vinyl ester, polyurethane, silicone, polyamide and polyamide-imide.

Exercise

A Multiple Choice Type Questions(MCQ) (Each Question carries 1 mark)

1. Who is the father of polymers?
   a. Hermann Staudinger
   b. Marie Curie
   c. Alexander von Humboldt
   d. Vladimir Komarov

2. DNA is
   a. synthetic polymer
   b. natural polymer
   c. none of the above
   c. both a and b

3. The word "polymer" was introduced by the Swedish chemist
   a. Paul John Flory
   b. Hermann Staudinger
   c. J. J. Berzelius
   d. both a and b

4. The first plastic material was based on
   a. acetate
   b. bakelite
   c. epoxy
   d. nitrocellulose

5. The first synthetic thermoset polymer is
   a. phenol-formaldehyde or bakelite
   b. nitrocellulose
   c. epoxy polymer
   d. amino polymer

6. Name of the Scientist/s got the Nobel Prize for conducting polymer study
   a. John Bennett Fenn, Koichi Tanaka, and Kurt Wüthrich
   b. Paul John Flory
   c. Alan J. Heeger, Alan MacDiarmid, and Hideki Shirakawa
   d. Giulio Natta and Karl Ziegler

7. Polymers are classified into two types: naturally occurring and synthetic on the basis of
   a. Tacticity
   b. Molecular Forces
   c. Crystallinity
   d. Sources of polymer

8. Which is synthetic organic polymer one of the following is?
   a. polyvinyl alcohol
   b. natural rubber
   c. natural silk
   c. cellulose

9. The Molecular forces present in polymers are
   a. Only van der Waals forces
   b. van der Waals forces, hydrogen bonding and dipole-dipole interactions
   c. Only hydrogen bonding
   d. Only dipole-dipole interactions

10. Which rubber is a natural polymer?

    a. vulcanized rubber

    b. neoprene, Buna rubbers, and butyl rubbers

    c. Styrene-butadiene rubber

    d. None of these

B Short Answer Type Questions (Each question carry either 2 or 3 marks)

Write short notes on the following

1. Nobel laureate in polymer chemistry

2. Classification of polymers

3. Draw different forms of polymer based on tacticity

4. Molecular forces in polymers

5. Elastomers

6. Fibres

7. Thermoplastic polymers

8. Thermosetting polymers

9. Texture of polymers

C Long Answer Type Questions (Each question carry 5 marks)

1. History of polymers

2. Classification of polymers with example

3. Molecular forces and chemical bonding in polymers

4. Copolymers

5. Texture of polymers based on macroscopically (using naked eye) and microscopically (using electron microscope or optical microscope)

6. Give examples of (a) commodity plastics, (b) engineering plastics, (c) elastomers, (d) transparent plastics, (e) thermoset resins for composites.

Answers

A

1. a	**2.** b	**3.** c	**4.** d	**5.** a	**6.** c	**7.** d	**8.** a	**9.** b	**10.** d

2

Functionality and its Importance

2.1 Functionality of Monomers

We know that the basic unit of polymer is monomer. In polymer, each monomer is attached (except terminal monomers) with two or more monomers. Thus, a monomer should have at least two bonding sites or functional groups or reactive sites to form a polymer. A monomer is called mono-functional, bi-functional or poly-functional depending on the number of bonding sites or functional groups or reactive sites present in the monomer.

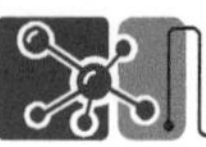
Mono-functional monomer

The monomer molecule is called mono-functional when it has one bonding site or functional group or reactive site. It may also be called unifunctional.

Example: CH_3NH_2, CH_3OH, CH_3COOH.

Bi-functional monomer

The monomer molecule is called bi-functional when it has two bonding sites or functional groups or reactive sites. It may also be called difunctional.

Example: $H_2C{=}CH_2$, $HOCH_2CH_2OH$, $HOOC(CH_2)_2COOH$, $H_2N(CH_2)_2NH_2$.

Poly-functional monomer

The monomer molecule is called poly-functional when it has more than two bonding sites or functional groups or reactive sites.

Example: $HOCH_2CH(OH)CH_2OH$, 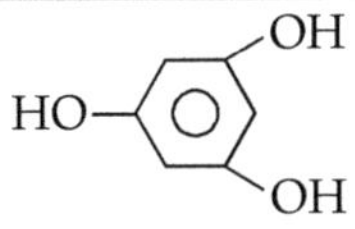

, 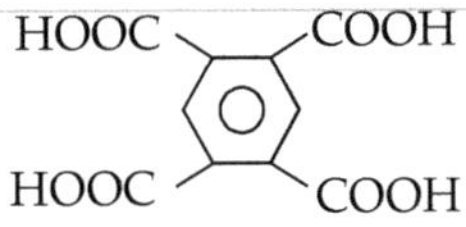

2.2 Degree of Polymerization

Degree of polymerization is defined as the number of monomer units present in the polymer. Mathematically it can be expressed as

$$\text{Degree of polymerization } (P) = \frac{\text{molar mass of polymer}}{\text{molar mass of monomer}}$$

Mathematically if we consider that, in a polymer there are 5000 monomer units and the molar mass of monomer is M. Then

$$\text{Degree of polymerization } (P) = \frac{5000\,\text{M}}{\text{M}} = 5000$$

2.3 Extent of Reaction

Extent of reaction in a polymerization reaction can be defined as the number of monomer units taking part in polymer formation divided by the total number of monomers taken. Mathematically, extent of reaction (p) can expressed as

$$p = \frac{\text{number of monomer units taking part in polymer formation}}{\text{total number of monomers taken}}$$

Before reaction starts the value of extent of reaction is 0 and at the end the value is 1. Suppose 5000 (N_0) monomer units are taken for polymerization reaction, after time t, 4000 (N_t) monomer units are present in the polymerization reaction. Thus, the number of monomer units taking part in polymer formation reaction is $(N_0 - N_t) = 5000 - 4000 = 1000$. Then

$$\text{Extent of reaction } (p) = \frac{N_0 - N_t}{N_0} = \frac{1000}{5000} = 0.2$$

By simplifying the equation $p = \frac{N_0 - N_t}{N_0} = \frac{N_0}{N_0} - \frac{N_t}{N_0} = 1 - \frac{N_t}{N_0}$

$$\frac{N_t}{N_0} = 1 - p \text{ or, } \frac{N_0}{N_t} = \frac{1}{1-p}$$

Again the number-average value of the degree of polymerization ($\overline{X}_n$) can be written as

$$\overline{X}_n = \frac{N_0}{N_t} = \frac{1}{1-p} \quad \text{or,} \quad \overline{X}_n = \frac{1}{1-p}$$

This equation is called the **Carothers equation.**

But Carothers gives the expression by combining the three terms namely extent of reaction, average degree of polymerization and average functionality. But in the above equation there is no average functionality term. So, this term should be incorporated in the above equation.

We assumed that there are N_0 molecules. If the molecule is bi-functional then the total number of functional groups will be $2 \times N_0$. Again, for the tri-functional molecule total number of functional groups will be $3 \times N_0$. In general, if the functionality is f then the total number of functional groups will be $f\,N_0$. Let each molecule lose one functional group as reaction proceeds. Thus two groups will be lost during polymerization (example, hydroxyl carboxylic acid polymerization). Thus total functional groups take part in the polymerization process will be $2(N_0 - N_t)$.

Hence $p = \dfrac{2}{f}\left(1 - \dfrac{N_t}{N_0}\right)$

Putting the value of $\overline{X}_n = \dfrac{N_0}{N_t}$ in the above equation $p = \dfrac{2}{f}\left(1 - \dfrac{1}{\overline{X}_n}\right)$

This equation is also called the **Carothers equation**.

When we put stoichiometric ratio of reactants (r = number of functional groups of A/ number of functional groups of B) the equation becomes

$$\overline{X}_n = \frac{1+r}{1+r-2rp}$$

1. **What do you understand by the functionality of monomers for polymer synthesis?**

Ans. In polymer a single molecule or different molecules are present as repeated units. One molecule can combine with another molecule if it has at least one active site or functional group. In polymer, each monomer is attached (except terminal monomers) with two or more monomers. Thus, a monomer should have at least two bonding sites or functional groups or reactive sites to form a polymer. On the basis of the number of active site or functional group present in monomer, monomer is called mono-functional monomer, bi-functional monomer and poly-functional monomer.

Mono-functional monomer : The monomer molecule is called mono-functional when it has one bonding site or functional group or reactive site. It may also be called unifunctional.

Example: CH_3NH_2, CH_3OH, CH_3COOH.

Bi-functional monomer: The monomer molecule is called bi-functional when it has two bonding sites or functional groups or reactive sites. It may also be called difunctional.

Example: $H_2C{=}CH_2$, $HOCH_2CH_2OH$, $HOOC(CH_2)_2COOH$, $H_2N(CH_2)_2NH_2$.

Poly-functional monomer : The monomer molecule is called poly-functional when it has more than two bonding sites or functional groups or reactive sites.

Example: $HOCH_2CH(OH)CH_2OH$, [structure: benzene ring with two OH groups], [structure: benzene ring with four COOH groups]

When the monomer has a π bond, it forms polymer by addition polymerization method. When the monomer has $-COOH$, $-NH_2$ and $-OH$ functional groups and small molecules are eliminated during the polymerization process, it follows the condensation polymerization method.

2. **What do you understand by the degree of polymerization and the extent of reaction.**

Ans. **Degree of polymerization :** Degree of polymerization is defined as the number of monomer units present in the polymer. Mathematically it can express as

$$\text{degree of polymerization } (P) = \frac{\text{molar mass of polymer}}{\text{molar mass of monomer}}$$

Extent of reaction : Extent of reaction in a polymerization reaction can be defined as the number of monomer units taking part in polymer formation divided by the total number of monomers taken. Mathematically extent of reaction (p) can be expressed as

$$= \frac{\text{number of monomer units taking part in polymer formation}}{\text{total number of monomers taken}}$$

3. **Show that for bifunctional monomer Carothers equation is** $\overline{X}_n = \dfrac{1}{1-p}$

Ans. We know that $p = \dfrac{2}{f}\left(1 - \dfrac{1}{\overline{X}_n}\right)$

Where p = extent of reaction

$\overline{X}_n$ = the number-average value of the degree of polymerization

f = functionality of monomer

for bifunctional monomer $f = 2$

then the equation become

$$p = \frac{2}{2}\left(1 - \frac{1}{\overline{X}_n}\right)$$

or, $p = 1 - \dfrac{1}{\overline{X}_n}$ or, $\overline{X}_n = \dfrac{1}{1-p}$

4. What is the physical significance of the value $p = 0$ and $p = 1$, where p = extent of reaction.

Ans. The value for extent of reaction is zero, signifies that the reaction is not started. When the reaction is completed, the value for the extent of reaction is one.

5. Show that $\overline{X}_n = \dfrac{1+r}{1+r-2rp}$

Ans. We know, average degree of polymerization $(\overline{X}_n)$ can be written as

$$\overline{X}_n = \frac{N_0}{N_t}$$

Where N_0 and N_t are the number of molecules present initially and number of unreacted molecules respectively.

extent of reaction $(p) = \dfrac{N_0 - N_t}{N_0}$

r = number of functional groups of A / number of functional groups of B, mathematically,

$$r = \frac{N_a}{N_b} \quad \text{or,} \quad N_a = rN_b$$

The total initial molecules are expressed in terms of functional group as

$$N_0 = \frac{N_a + N_b}{2}$$

or, N_a can be replaced by N_b

$$N_0 = \frac{rN_b + N_b}{2} \quad \text{or, } N_0 = \frac{N_b(r+1)}{2}$$

Now we have to calculate number of unreacted molecules in terms of number of functional groups of A or B. Consider the extent of reaction for this calculation

$$\text{unreacted } A = N_a - pN_a = rN_b - prN_b$$

and *unreacted* $B = N_b - pN_b$

as the extent of reaction are similar with functional group, we can write $pN_b = pN_a$

thus *unreacted* $B = N_b - pN_b = N_b - pN_a = N_b - prN_b$

total unreacted molecules are $(N_t) = (\textit{unreacted } A + \textit{unreacted } B)/2$ [considering the functionality 2]

[if a molecule is bifunctional then when we count a molecule, it is one but when we count functionality is 2. Thus, for the conversion of functionality to molecule it is divided by 2]

$$\text{or } N_t = \frac{rN_b - prN_b + N_b - prN_b}{2} = \frac{N_b(1+r-2pr)}{2}$$

again, we know

$$\overline{X}_n = \frac{N_0}{N_t}$$

$$\text{or, } \overline{X}_n = \frac{\dfrac{N_b(r+1)}{2}}{\dfrac{N_b(1+r-2pr)}{2}}$$

$$\text{or, } \overline{X}_n = \frac{(r+1)}{(1+r-2pr)}$$

Exercise

A **Multiple Choice Type Questions(MCQ) (Each Question carries 1 mark)**

1. Functionality of monomers will depend on
   a. number of bonding sites
   b. number of elements present
   c. the ratio of elements and bonds
   d. none of these

2. Which is Mono-functional monomer?
   a. oxalic acid
   b. acetic acid
   c. phthalic acid
   d. benzene

3. Which is bi-functional monomer?
   a. sodium hydroxide
   b. oxalic acid
   c. acetic acid
   d. benzene

4. Which is poly-functional monomer?
   a. acetate
   b. bakelite
   c. epoxy
   d. Phloroglucinol

5. Degree of polymerization is the ratio of
   a. molar mass of polymer and molar mass of initiator
   b. molar mass of polymer and molar mass of monomer
   c. molar mass of monomer and molar mass of polymer
   d. none of these

6. Carothers gives the expression by combining the terms
   a. extent of reaction, average degree of polymerization and average functionality
   b. extent of reaction, average degree of polymerization
   c. average degree of polymerization and average functionality
   d. none of these

B **Short Answer Type Questions (Each question carry either 2 or 3 marks)**
Write short notes on the following

1. Mono-functional monomer

2. Bi-functional monomer

3. Poly-functional monomer

4. Degree of polymerization

5. Extent of reaction

6. Carothers equation

C Long Answer Type Questions (Each question carry 5 marks)

1. What do you understand by the functionality of monomers?

2. Show that $\overline{X}_n = \dfrac{1+r}{1+r-2rp}$

Answers

A

1. a	**2.** b	**3.** c	**4.** d	**5.** b	**6.** b

3

Kinetics of Polymerization

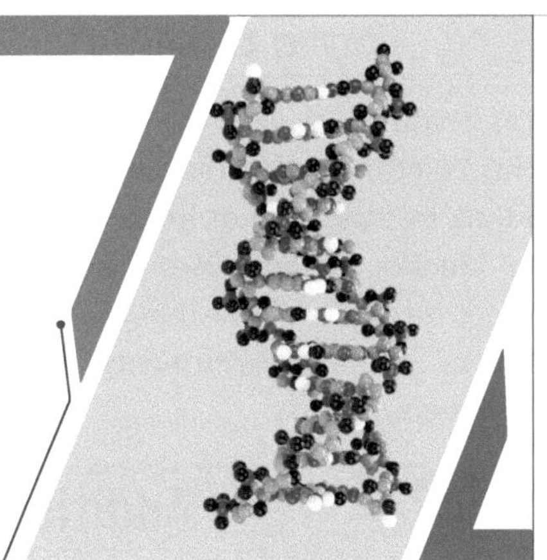

Polymerization is the process by which monomer molecules bind each other in a chemical reaction to form polymer chains or networks. There are generally two types of polymerization processes namely addition polymerization (chain growth reaction) and condensation polymerization (step growth reaction). But later on, polymerization was divided into four types namely addition polymerization, condensation polymerization, coordination polymerization and copolymerization. Each polymerization process occurs through a particular mechanism. For example, in addition polymerization, the chain growth will take place through free radicals or ionic species formation. Ionic species can be two types cationic and anionic. Free radical polymerization is most commonly occurring polymerization rather than cationic and anionic polymerization. Now each polymerization process will be discussed one by one.

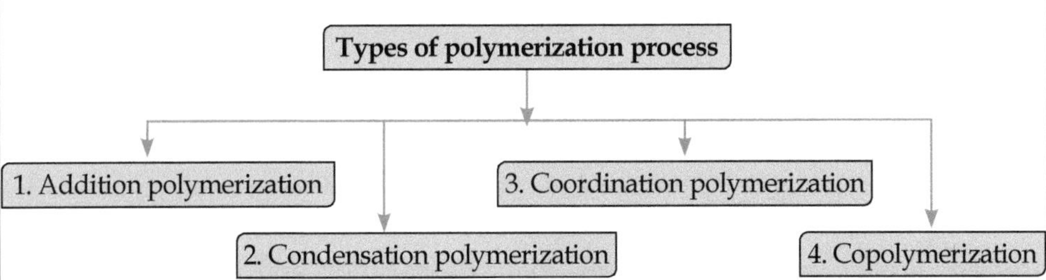

As the reaction proceeds, the molecular weight of living polymer chains (roughly assume that the polymer chains which can further grows) increases rapidly. For linear polymer, chains grow linearly but for cross-linked or network polymer, several linear chains become linked together.

To study the kinetics of ester hydrolysis, the reaction is stopped or slowed down by adding ice water in the mixture. But in polymer synthesis reaction (i) the process is very fast (polymer gets prepared before you stop reaction)

(ii) if you try to stop the reaction in any stage you will come up with oligomers (iii) You cannot stop the reaction by lowering temperature or any other procedure. Many instrumental techniques are used to examine the kinetics of polymerization or kinetics of thermoset cure namely in-situ FTIR (Fourier transform infrared spectroscopy), UV-Visible spectra, DSC (Differential scanning calorimetry) and others. All the instruments have limitations and are applicable for definite polymers.

3.1 Addition Polymerization

Addition polymerization is also called **chain growth polymerization**. This type of polymerization proceeds through the formation of free radicals or ionic species (cationic or anionic). Addition polymerization can be subdivided into three categories namely free radical polymerization, cationic polymerization and anionic polymerization.

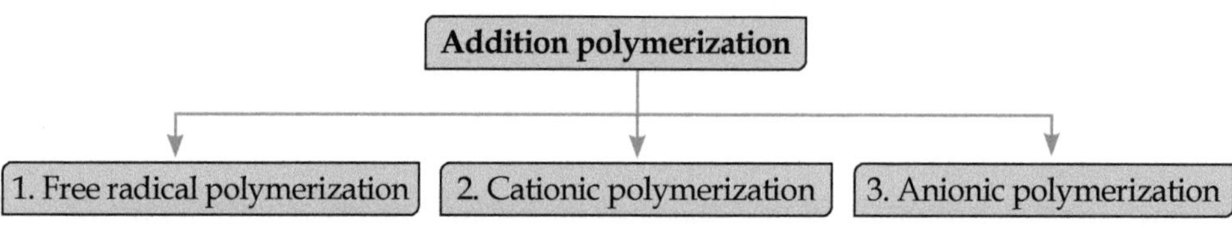

In this polymerization process monomer (same or mixed monomer) and catalyst (free radical or ion generating molecule) are taken. The reactions proceed through chain initiation, chain propagation and chain termination. All the monomer molecules take part in this polymerization.

Free radical polymerization

In this polymerization monomer and catalyst (free radical generating molecule) are taken. Initially radical is generated in the chain initiation step. The reactions proceed through radical generation, chain propagation and chain termination. Generally alkenes and their derivatives are used as monomers and benzoyl peroxide, acetyl peroxide etc are used as catalysts.

Few important catalyst compounds & their radicals.

Compound	Radical ($\dot{R}$)
benzoyl peroxide	benzoyloxy radical → phenyl radical

Compound	Radical ($\dot{R}$)
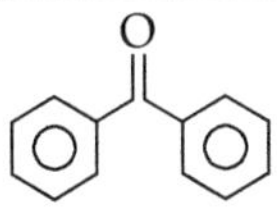 benzophenone	 benzoyl radical + phenyl radical
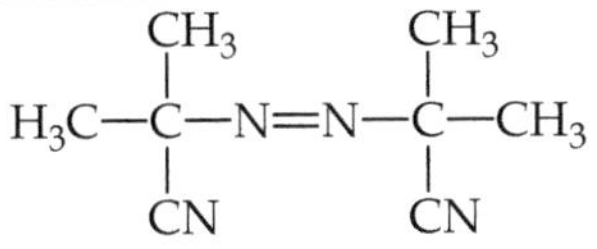 $H_3C-\underset{\underset{CN}{\mid}}{\overset{\overset{CH_3}{\mid}}{C}}-N=N-\underset{\underset{CN}{\mid}}{\overset{\overset{CH_3}{\mid}}{C}}-CH_3$ 2,2'-azo-bis-isobutyronitrile (AIBN)	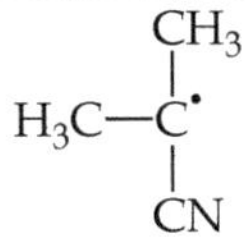 $H_3C-\underset{\underset{CN}{\mid}}{\overset{\overset{CH_3}{\mid}}{\dot{C}}}$ 2-cyano-2-propyl radical
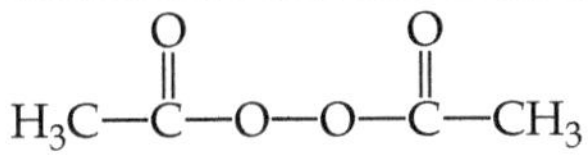 $H_3C-\overset{\overset{O}{\parallel}}{C}-O-O-\overset{\overset{O}{\parallel}}{C}-CH_3$ Acetyl peroxide	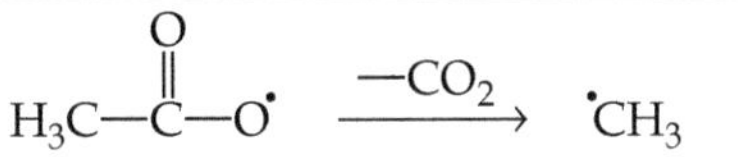 $H_3C-\overset{\overset{O}{\parallel}}{C}-\dot{O} \xrightarrow{-CO_2} \dot{C}H_3$ acetyloxy radical methyl radical

Example: Methyl methacrylate, styrene, vinyl chloride and acrylonitrile monomers follow the free radical polymerization mechanism.

Mechanism and kinetics: There are three primary steps *i.e.*, **initiation, propagation,** and **termination** in free radical polymerization process. In the initiation step radicals are formed, in the propagation step molecular weight of living polymer increases and in termination step living polymer becomes dead and gets the polymer.

i) **Initiation :** It is the first step of radical polymerization. In this step an active centre is created from which a polymer chain is started to grow. Thermal or photochemical decomposition of organic peroxides, hydroperoxides, azo or diazo compounds, high-energy radiation, oxidation-reduction (redox) reactions and electrochemical reactions are the ways for free radical creation.

initiator $(I) \rightarrow$ radical $(\dot{R})$ or radicals $(n\dot{R})$

radical $(\dot{R})$ + monomer $(M) \rightarrow$ monomer radical $(R\dot{M})$

Thus the rate equation is

$$R_i = \frac{d[R\overset{\infty}{\dot{M}}]}{dt} = nfk_i[I]$$

For homolytic decomposition of initiator, $n = 2$

$$R_i = \frac{d[R\overset{\infty}{\dot{M}}]}{dt} = 2fk_i[I]$$

where k_i is rate constant and f is fraction of monomers taking part in this step (generally the value of f is 0.3 to 0.8).

Actually, the initiation process proceeds through two sub-steps. In the first step initiator molecules create one or more than one radical depending on which type of catalyst is used. Then this radical will react with a monomer molecule giving a monomer radical. It was found that the formation of radical from initiator is the slowest step between these two steps.

ii) **Propagation :** It is the second step of radical polymerization. In this step the growth of a polymer chain occurs by successive addition of monomers.

$$R\dot{M} + M \rightarrow \dot{M}_2$$
$$\dot{M}_2 + M \rightarrow \dot{M}_3$$
$$\dot{M}_n + M \rightarrow \dot{M}_{n+1}$$

Thus the rate equation is

$$R_p = \frac{-d[M]}{dt} = k_p[M][\dot{M}]$$

where k_p is rate constant.

Radicals attack the loosely bound pi bond of the alkene forming another free radical and the process continues until stop in the termination step.

iii) **Termination :** It is the final step of radical polymerization. In this step the growth of a polymer chain stops and obtains the final polymer. Combination of two living polymer chains, radical disproportionation and Combination of a living polymer chain with an initiator radical or impurities or inhibitors will stop the polymerization process that is polymer chains get terminated.

$$\dot{M}_x + \dot{M}_y \rightarrow M_{x+y}$$
$$\dot{M}_x + \dot{M}_y \rightarrow M_x + M_y$$
$$\dot{M}_x + \dot{R} \rightarrow M_xR$$

Thus the rate equation is

$$R_t = \frac{-d[\dot{M}]}{dt} = k_t[\dot{M}][x]$$

where k_t is rate constant and $[x]$ is either another living polymer chain or initiator or inhibitor. When the reaction stops by other living polymer chain, rate equation will be

$$R_t = \frac{-d[\dot{M}]}{dt} = 2k_t[\dot{M}]^2$$

Using steady-state approximation,

Rate of initiation (R_i) = Rate of termination (R_t)

$$2fk_i[I] = 2k_t[\dot{M}]^2$$

$$[\dot{M}] = \sqrt{\frac{fk_i[I]}{k_t}}$$

Then the rate of polymerization

$$R_p = \frac{-d[\dot{M}]}{dt} = k_p[M][\dot{M}] = k_p[M]\sqrt{\frac{fk_i[I]}{k_t}}$$

$$R_p = k_p\left(\frac{fk_i}{k_t}\right)^{\frac{1}{2}}[M][I]^{\frac{1}{2}}$$

Thus, the rate of polymerization is depending on the rate constants of these three steps and concentration of monomer and initiator.

- Example of free radical polymerization of ethylene by benzoyl peroxide

Initiation :

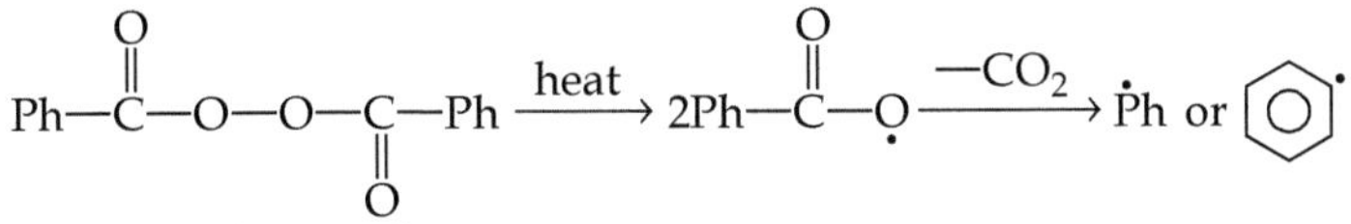

benzoyl peroxide

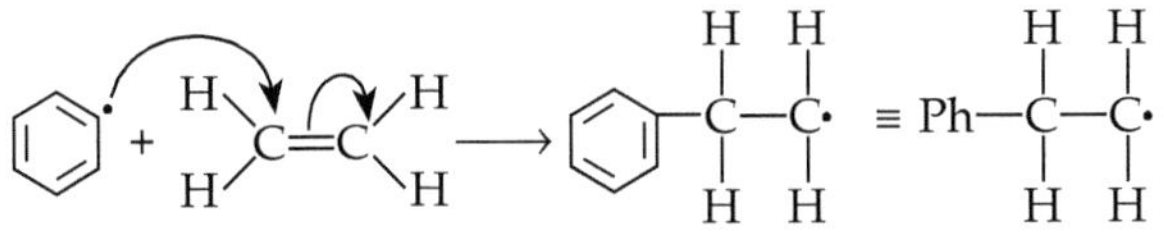

Propagation :

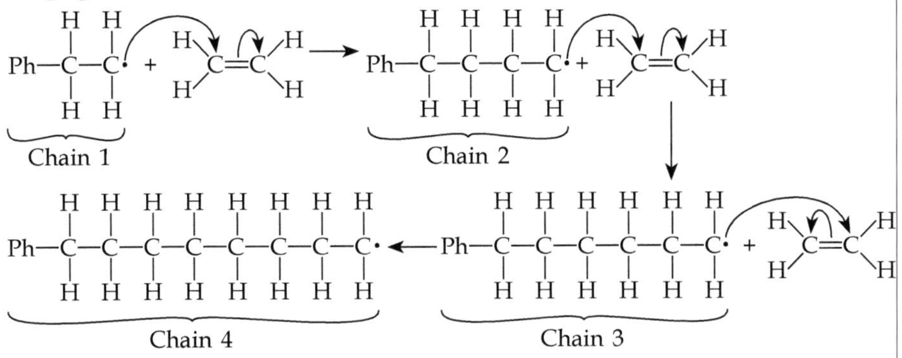

Chain 1   Chain 2

Chain 4   Chain 3

In propagation step many chains (Chain 1, Chain 2, ... Chain n) are formed.

Termination :

In termination step any two chains from chain 1 to chain n combined with each other. As example

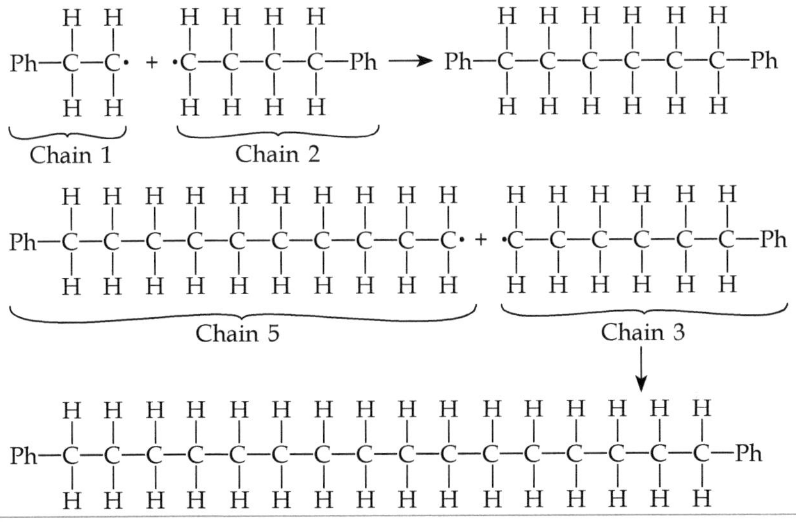

Chain 1   Chain 2

Chain 5   Chain 3

 ## Cationic polymerization

In this polymerization monomer and catalyst (cation generating molecule) are taken. Initially cation is generated in the chain initiation step. The reactions proceed through cation generation, chain propagation and chain termination. Generally, alkenes and their derivatives are used as monomers and Lewis acids BF_3, $SnCl_4$, $AlCl_3$ etc with H_2SO_4, HF are used as catalysts.

Example : Styrene, propylene and vinyl acetate monomers follow the cationic polymerization mechanism.

Mechanism and kinetics: Kinetics is similar like free radical polymerization but mechanism is little bit different. In the initiation step instead of radical, cation is created.

In initiation step

 ## Anionic polymerization

In this polymerization monomer and catalyst (anion generating molecule) are taken. Initially anion is generated in the chain initiation step. The reactions proceed through **anion generation, chain propagation** and **chain termination**. Generally, alkenes and their derivatives are used as monomers and potassium amide, *n*-butyl lithium etc are used as catalysts.

Examples: Methyl methacrylate, styrene, vinyl chloride and acrylonitrile monomers follow the anionic polymerization mechanism.

Mechanism and kinetics: Kinetics is similar to free radical polymerization but the mechanism is a little bit different. In the initiation step instead of radical, anion is created.

In initiation step

3.2 Condensation Polymerization

Condensation polymerization is also called **step growth polymerization** (as the polymer chains grow step by step). In addition polymerization the number of monomer is one or more but for condensation polymerization the number of monomer must be two or more. The monomers used in condensation polymerization can be bi-functional or multifunctional. In this process catalyst may or may not be used. The polymer chains remain active even after polymerization is completed as the terminal monomer always has mono-functionality. For understanding you may assume A and B are the two monomers going to polymerize using condensation mechanism. As the monomer must be bi-functional so let A and B both have two hands. When A react with B dimer is formed. Remember that the nature of hands of A and B are different. A can only hold B's hand or vice-versa. Thus, dimer will form trimer and then tetramer and the process will continue. Diol, diamide, dicarboxylic acid etc. are the examples of the monomers.

Example of condensation polymers : polyesters, polyamides, polyurethanes, Polycarbonates, Bakelite, DNA (deoxyribonucleic acid) and others.

Mechanism and kinetics : This type of polymerization process performs either using catalyst or without catalyst. According to Flory's equal reactivity principle, the formation of dimers, trimers and all other steps have equal rate constants. This assumption makes the complicated kinetics to be very simple one. An example of condensation polymerization is the formation of polyester from diol and dicarboxylic acid. let the concentration of both diol and dicarboxylic acid are the same and that is C.

$$\text{diol} + \text{dicarboxylic acid} \rightarrow \text{polyester} + H_2O$$

$$HO-(CH_2)_m-OH + HOOC-(CH_2)_n-COOH \rightarrow$$

$$HO-(CH_2)_m-OOC-(CH_2)_n-COOH + H_2O$$

$$\text{Rate} = \frac{d[\text{polyester}]}{dt} = -\frac{d[\text{diol}]}{dt} = -\frac{d[\text{dicarboxylic acid}]}{dt}$$

Then Rate $(R) = -\dfrac{d[C]}{dt} = k[C]^2$

where k is rate constant.

On rearrangement, $-\dfrac{d[C]}{[C]^2} = kdt$

On integration, $\displaystyle\int_{C_0}^{C_t} -\frac{d[C]}{[C]^2} = \int_{t=0}^{t=t} kdt$

$$\frac{1}{[C]_t} - \frac{1}{[C]_0} = kt$$

If we measure the inverse of monomer concentration (it can be measured by titrating unreacted monomers in the mixture) with time, we can determine k. The slope will give an inverse of monomer's initial concentration.

Mechanism :

$$HO-(CH_2)_m-O\boxed{H + HO}OC-(CH_2)_n-COOH$$

$$\downarrow -H_2O$$

$$HO-(CH_2)_m-O-\overset{\overset{\textstyle O}{\|}}{C}-(CH_2)_n-COOH$$

$\sim\sim$COOH will attack at this side $\sim\sim$OH will attack at this side

3.3 Coordination Polymerization

Karl Ziegler and Giulio Natta invented the coordination polymerization and were awarded the Nobel Prize in chemistry in 1963. Generally non-polar monomers like 1-alkene, cycloalkenes, dienes, and alkynes are used for this type of polymerization. The catalyst acts either in a homogeneous medium or in a heterogeneous medium. The compositions of catalyst in homogeneous medium is transition metal metallocene and aluminum alkyl halide whereas for heterogeneous medium is transition metal compound ($TiCl_4$, VCl_3, $ZrCl_4$ etc.) and organometallic compounds ($AlEt_3$, $AlEt_2Cl$ etc.).

Examples of coordination polymers : polyethylene, polystyrene, Polyvinyl chloride, polyacrylonitrile etc.

Mechanism and kinetics : Kinetics is similar to free radical polymerization but the mechanism is a little bit different. Till now the mechanism of coordination polymerization is not clear but is assumed that the transition metal forms the active site. When the polymerization proceeds through one metal active site is termed monometallic mechanism and when proceed through two metal active sites is termed bimetallic mechanism.

Active site is formed by the adsorption of metal alkyl from solution

$$TiCl_4 + AlEt_3 \rightarrow TiEtCl_3 \text{ (active site)} + AlEt_2Cl$$

Schematically we can write

$$\text{▥} + MX \underset{\longleftarrow}{\overset{k_1}{\longrightarrow}} \text{▥} - XR$$

where ▥ is transition metal compound and XR is metal alkyl compound.

When monomer (M) is adsorbed by transition metal

$$\text{▥} + M \underset{\longleftarrow}{\overset{k_2}{\longrightarrow}} M - \text{▥}$$

Thus in the initiation step

$$M - \text{▥} - XR \underset{\longleftarrow}{\overset{k_1}{\longrightarrow}} \text{▥} - X - M - R$$

In the propagation step

$$M - \boxed{} - X - M - R \xrightleftharpoons{k_p} \boxed{} - X - (M)_n - R$$

In the termination step

$$M - \boxed{} - X - (M)_n - R \xrightleftharpoons{k_t} \boxed{} - X - M - R + (M)_n$$

3.4 Copolymerization

Copolymer is formed by combining more than one monomer and the process is termed as copolymerization. When two or more monomers are combined to form copolymer is called bipolymer, terpolymer, quaterpolymer and so on periodically. The classifications of copolymers are discussed before.

Examples of copolymers: Polyethylene-vinyl acetate (PEVA), nitrile rubber, nylon 6, nylon 66 etc.

Mechanism and kinetics : Let M_1 and M_2 are the two monomers those form copolymer by free radical polymerization path. It is assumed that the polymerization process occurs through two initiations, two terminations and four propagations steps. As either monomer M_1 or M_2 can create radical in the initiation step. The living polymer chains either have monomer M_1 radical or M_2 radical at the end of polymer chains which can combine either M_1 or M_2 monomer giving four possibilities in the propagation step. Again in the termination step the living polymer chains (either having monomer M_1 radical or M_2 radical at the end of polymer chains) giving two possibilities to terminate the polymerization.

In the initiation step

$$\text{Initiator } (I) \rightarrow \text{Radical } (\overset{\bullet}{R}) \text{ or Radicals } (n\overset{\bullet}{R})$$

$$\overset{\bullet}{R} + M_1 \rightarrow R\overset{\bullet}{M}_1$$

$$\overset{\bullet}{R} + M_2 \rightarrow R\overset{\bullet}{M}_2$$

In the propagation step

$$R\overset{\bullet}{M}_1 + M_1 \xrightarrow{k_{11}} RM_1\overset{\bullet}{M}_1$$

$$R\overset{\bullet}{M}_1 + M_2 \xrightarrow{k_{12}} RM_1\overset{\bullet}{M}_2$$

$$R\overset{\bullet}{M}_2 + M_1 \xrightarrow{k_{21}} RM_2\overset{\bullet}{M}_1$$

$$R\overset{\bullet}{M}_2 + M_2 \xrightarrow{k_{22}} RM_2\overset{\bullet}{M}_2$$

Using penultimate effects (the reaction between monomer at chain end and the monomer added in the previous step is independent *i.e.*, $RM_1\overset{\bullet}{M}_1 + M_2$ is similar as $RM_2\overset{\bullet}{M}_1 + M_1$) the rates of these four propagation steps are

$$R_{11} = k_{11}[\overset{\bullet}{M}_1][M_1]$$

$$R_{12} = k_{12}[\overset{\bullet}{M}_1][M_2]$$

$$R_{21} = k_{21}[\overset{\bullet}{M_2}][M_1]$$

$$R_{22} = k_{22}[\overset{\bullet}{M_2}][M_2]$$

The rate of formation of living chain

$$\frac{d[\overset{\infty\bullet}{M_1}]}{dt} = -k_{12}[\overset{\bullet}{M_1}][M_2] + k_{21}[\overset{\bullet}{M_2}][M_1]$$

and $$\frac{d[\overset{\infty\bullet}{M_2}]}{dt} = -k_{21}[\overset{\bullet}{M_2}][M_1] + k_{12}[\overset{\bullet}{M_1}][M_2]$$

using steady state approximation

$$\frac{d[\overset{\infty\bullet}{M_1}]}{dt} = \frac{d[\overset{\infty\bullet}{M_2}]}{dt} = 0$$

i.e., $-k_{12}[\overset{\bullet}{M_1}][M_2] + k_{21}[\overset{\bullet}{M_2}][M_1] = -k_{21}[\overset{\bullet}{M_2}][M_1] + k_{12}[\overset{\bullet}{M_1}][M_2]$

Simplifying

$$2k_{12}[\overset{\bullet}{M_1}][M_2] = 2k_{21}[\overset{\bullet}{M_2}][M_1]$$

$$\frac{[\overset{\infty\bullet}{M_1}]}{[\overset{\infty\bullet}{M_2}]} = \frac{k_{21}[M_1]}{k_{12}[M_2]}$$

There are two types of mole fraction, one is in the reaction mixture and another type is in the copolymer preparation. Let f_1 and f_2 are the mole fraction of monomers in the reaction mixture and F_1 and F_2 are the mole fraction of monomers in the copolymer formed. As the reaction proceeds the amount of free monomer decreases and the amount of monomer in the copolymer increases. The f_1 and f_2 are fractional quantity thus

$$f_1 + f_2 = 1$$

Similarly, $F_1 + F_2 = 1$

The mole fraction of monomers present in the copolymer is

$$F_1 = \frac{d[M_1]}{dt} = -k_{11}[\overset{\bullet}{M_1}][M_1] - k_{21}[\overset{\bullet}{M_2}][M_1]$$

and $F_2 = \dfrac{d[M_2]}{dt} = -k_{12}[\overset{\bullet}{M_1}][M_2] - k_{22}[\overset{\bullet}{M_2}][M_2]$

The ratio of F_1 and F_2 is

$$\frac{F_1}{F_2} = \frac{-k_{11}[\overset{\bullet}{M_1}][M_1] - k_{21}[\overset{\bullet}{M_2}][M_1]}{-k_{12}[\overset{\bullet}{M_1}][M_2] - k_{22}[\overset{\bullet}{M_2}][M_2]}$$

Multiplying the r.h.s. by $(-1/[\overset{\bullet}{M_2}])$ on the both numerator and denominator

$$\frac{F_1}{F_2} = \frac{k_{11}[\overset{\bullet}{M_1}/\overset{\bullet}{M_2}][M_1] - k_{21}[M_1]}{k_{12}[\overset{\bullet}{M_1}/\overset{\bullet}{M_2}][M_2] - k_{22}[M_2]}$$

But we know from steady state approximation

$$\frac{[\overset{\infty\bullet}{M_1}]}{[\overset{\infty\bullet}{M_2}]} = \frac{k_{21}[M_1]}{k_{12}[M_2]}$$

So, $\dfrac{F_1}{F_2} = \dfrac{\dfrac{k_{11}k_{21}}{k_{12}}\dfrac{[M_1]^2}{[M_2]}+k_{21}[M_1]}{\dfrac{k_{12}k_{21}}{k_{12}}[M_1]+k_{22}[M_2]}$

Multiplying the r.h.s. by $1/k_{21}[M_2]$ on the both numerator and denominator

$$\dfrac{F_1}{F_2} = \dfrac{\dfrac{k_{11}}{k_{12}}\dfrac{[M_1]^2}{[M_2]^2}+\dfrac{[M_1]}{[M_2]}}{\dfrac{[M_1]}{[M_2]}+\dfrac{k_{22}}{k_{21}}}$$

Let r_1 and r_2 are the reactivity ratios defined as

$$r_1 = \dfrac{k_{11}}{k_{12}} \text{ and } r_2 = \dfrac{k_{22}}{k_{21}}$$

Then $\dfrac{F_1}{F_2} = \dfrac{r_1\dfrac{[M_1]^2}{[M_2]^2}+\dfrac{[M_1]}{[M_2]}}{\dfrac{[M_1]}{[M_2]}+r_2}$

We know $\dfrac{f_1}{f_2} = \dfrac{[M_1]}{[M_2]}$

then, $\dfrac{F_1}{F_2} = \dfrac{r_1\dfrac{[f_1]^2}{[f_2]^2}+\dfrac{[f_1]}{[f_2]}}{\dfrac{[f_1]}{[f_2]}+r_2} = \dfrac{r_1\dfrac{[f_1]}{[f_2]}+1}{1+r_2\dfrac{[f_2]}{[f_1]}}$

We can also express the copolymer equation in terms of mole fraction of one monomer in copolymer and feed as

$$\dfrac{F_1}{1-F_1} = \dfrac{r_1\dfrac{[f_1]}{[1-f_1]}+1}{1+r_2\dfrac{[1-f_1]}{[f_1]}}$$

From this equation we can conclude

i. Alternating copolymer will form when $r_1 = r_2 = 0$.

ii. Random copolymer will form when $r_1 = r_2 = 1$.

It can also be concluded that which monomer favours reaction and which does not.

i. Both radicals favour to react with the M_1 monomer when $r_1 > 1$ and $r_2 < 1$.

ii. Both radicals favour to react with the M_2 monomer when $r_1 < 1$ and $r_2 > 1$.

iii. Both radicals favour to react with their own monomer when $r_1 > 1$ and $r_2 > 1$.

iv. Neither radicals favour to react with their own monomer when $r_1 < 1$ and $r_2 < 1$.

3.5 Polymerization Techniques

Researchers are endlessly developing novel polymerization techniques. Though it is a continuous process but there are four well known polymerization techniques. These four techniques are

1. Solution polymerization techniques
2. Bulk polymerization techniques
3. Emulsion polymerization techniques
4. Suspension polymerization techniques

The examples are presented in the following table

Polymerization technique	Polymers
Solution	Polyacrylonitrile, Polypropylene, Polystyrene, Polyisoprene, Polyethylene
Bulk	Polyamide, Polycarbonate, Poly(methyl methacrylate), Nylon 6
Emulsion	Neoprene, Nitrile rubber, Polyvinyl fluoride, Polyvinyl acetate
Suspension	Polyvinyl chloride, Polysulphide, Poly(styrene-acrylonitrile), Poly(methyl methacrylate)

Solution polymerization

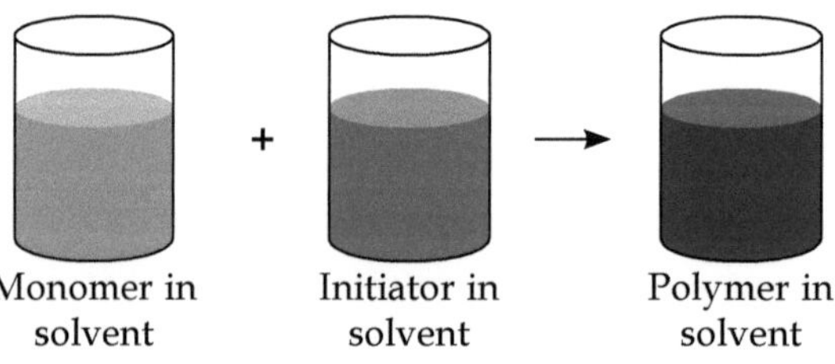

Generally in solution polymerization a solvent is chosen in such a way that both monomer and initiator are dissolved in the solvent without reacting. Then the two solutions are mixed together at a time or gradually with each other. The mixing procedure can be varied. The temperature of the reaction is a dominating factor to control the molecular weight of the polymer. Depending on the solubility of the polymer in this solvent it produces homogeneous polymer solution or heterogeneous polymer solution. By this method low molecular weight polymers are formed because of less monomer concentration in solution and chain gets terminated by solvent. The morphology of the polymer can be tuned by changing monomer concentration, initiator, solvent, temperature and using template in the solution.

Bulk polymerization

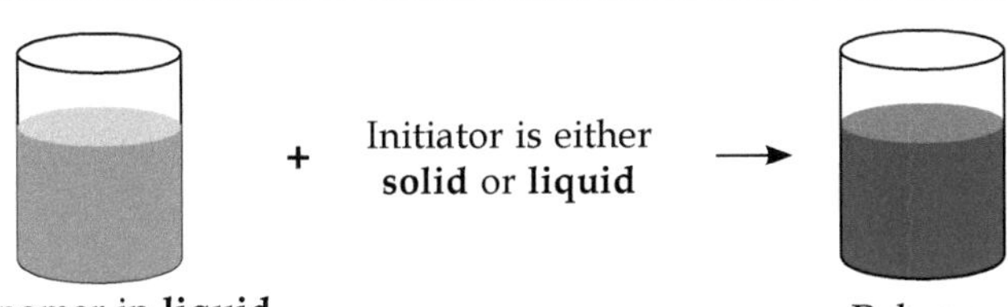

Monomer in **liquid** + Initiator is either **solid** or **liquid** → Polymer

In this polymerization process monomer is always in liquid state and the initiator is either in solid or in liquid state which is soluble in pure monomer. The reaction is initiated by thermally or photochemically. As there are no waste products or the final form of polymer does not need any purification, this process is environmentally friendly.

Emulsion polymerization

In this polymerization process **monomer, emulsifier, initiator and dispersing medium** are used. In particular cases modifiers are also used in the system. Water is normally used as a dispersing medium. The monomer is slightly soluble or insoluble in water but the initiator is soluble in water. Surfactant is used as an emulsifier which forms micelles around the monomer droplet. Surfactant may be cationic (quaternary ammonium salts, benzalkonium, benzethonium, methylbenzethonium, cetylpyridinium), anionic (ammonium lauryl sulfate, sodium laureth sulfate, sodium lauryl sarcosinate), non-ionic [poly(ethylene oxide), poly(vinyl alcohol) and hydroxyethyl cellulose] and zwitterionic (betaines, lauryldimethylamine-N-oxide). The molecular weight achieved by this method is higher than bulk or solution polymerization

When only surfactant is in a dispersing medium it forms micelle above CMC (critical micelle concentration).

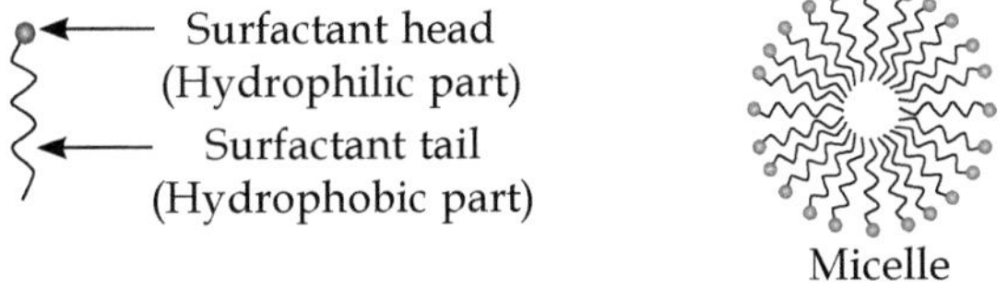

Slightly soluble or insoluble monomers are dispersed in water in presence of surfactant forming core-shell structure. Monomer droplet remains in the core and surfactant is in the shell.

Monomer molecules to monomer droplet

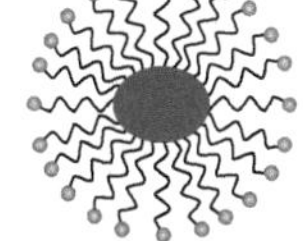

Monomer droplet in the micelle

After addition of initiator molecule, monomer molecules start polymerization

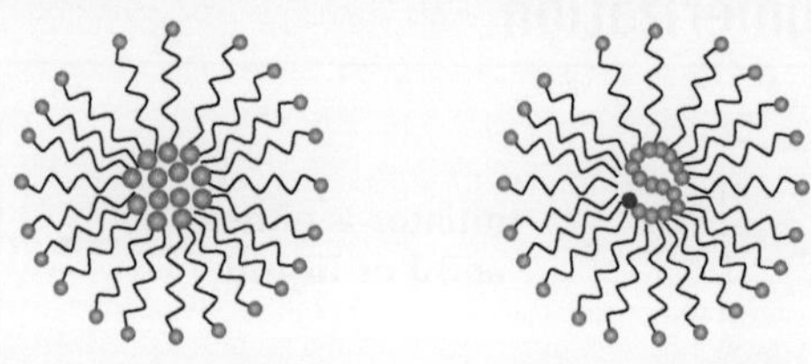

Micellar structure of surfactants can be broken by the addition of demulsifiers and then the living polymers combine with each other and polymerization stops.

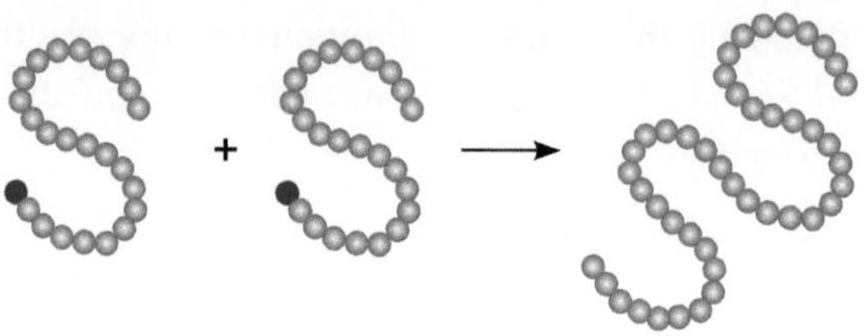

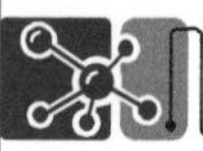 ## Suspension polymerization

This is a heterogeneous radical polymerization process where **monomer, stabilizing agent, initiator and dispersing medium** are used. Mechanical agitation is required for this process. Water is normally used as a dispersing medium. The monomer is slightly soluble or insoluble in water and the initiator is soluble in monomer. Stabilizing agent is used to stabilize monomer-initiator droplets in the dispersion. The initiators used are benzoyl peroxide, AIBN, diacetyl peroxide, lauroyl peroxide and others. Stabilizers may be surfactants or other compounds. The common stabilizers used are methyl and ethyl cellulose, PVA, alginate, casein, gelatin, $MgCO_3$, $CaCO_3$, and $CaSO_4$ and others. The polymers formed by this polymerization process have bead, pearl or granular like structures. Thus it is also termed as bead, pearl or granular polymerization.

Monomer will stay in either upper layer or lower layer when mixed with solvent.

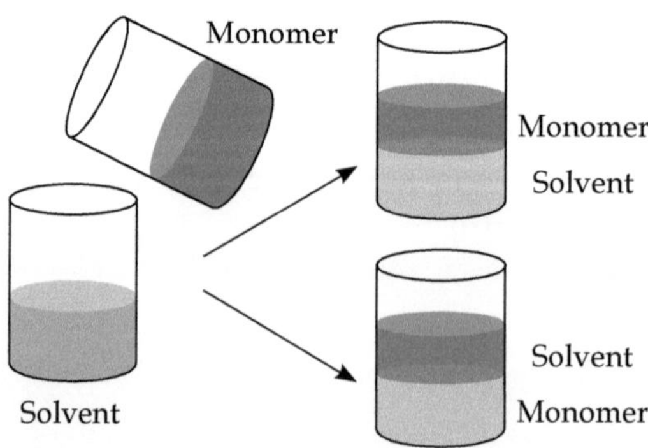

After addition of stabilizer with mechanical agitation it becomes dispersion.

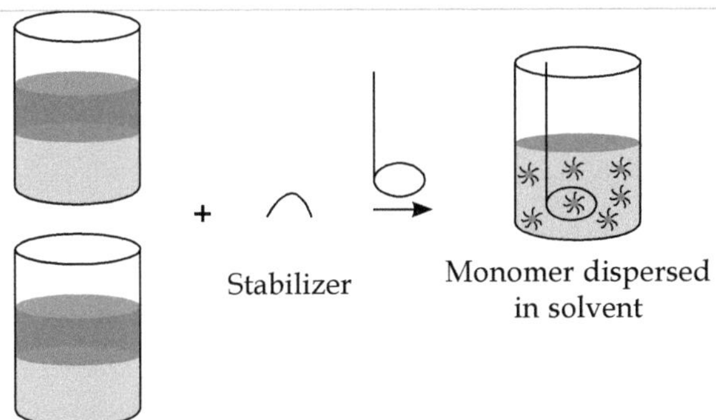

The initiator is soluble in monomer and hence the stabilizer stabilizes monomer incorporated with initiator. Dispersion behaves as a micro-reactor and polymerization starts in this micro-reactor. According to the shape of the micro-reactor the polymer formed beads, pearls or granules. This is shown by schematic presentation below.

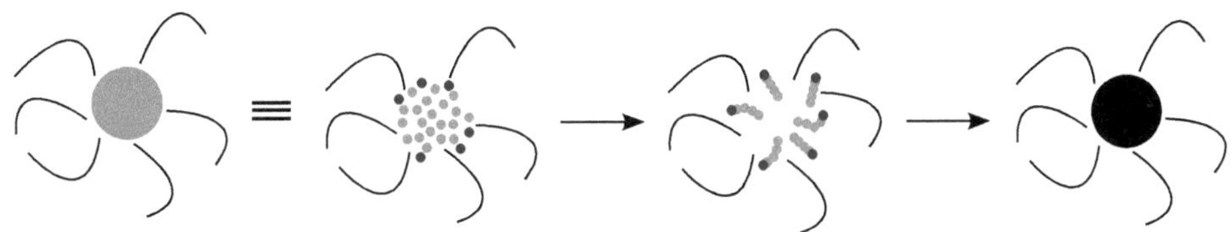

Finally formed blue polymer beads.

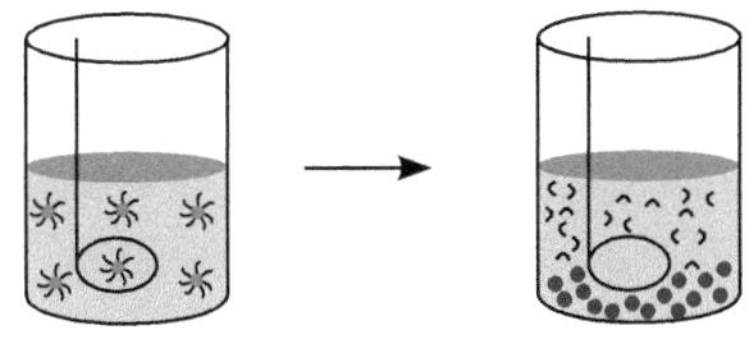

QUESTION & ANSWER

1. Describe the types of polymerization process.

Ans. Generally, monomer molecules combine with each other in a chemical reaction to form polymer chains (linear chain, branched chain or network). This process is called polymerization process. Till now there are four types of polymerization process namely addition polymerization, condensation polymerization, coordination polymerization and copolymerization.

▶ Addition polymerization is also called chain growth polymerization. This type of polymerization proceeds through the formation of free radicals or ionic species (cationic or anionic).

Example: Styrene, propylene and vinyl acetate.

▶ Condensation polymerization is also called step growth polymerization (as the polymer chains grow step by step). In addition polymerization the number of monomer is one or more but for condensation polymerization the number of monomer must be two or more. The monomers used in condensation polymerization can be bi-functional or multifunctional. In this process catalyst may or may not be used and small molecules elimination take place. The polymer chains remain active even after polymerization is completed as the terminal monomer always has mono-functionality.

Example: Polyesters, polyamides, polyurethanes, Polycarbonates, Bakelite and DNA (deoxyribonucleic acid).

▶ Generally non-polar monomers like 1-alkene, cycloalkenes, dienes, and alkynes are used for coordinationpolymerization. The catalyst acts either in a homogeneous medium or in a heterogeneous medium. The compositions of catalyst in homogeneous medium is transition metal metallocene and aluminium alkyl halide whereas for heterogeneous medium is transition metal compound ($TiCl_4$, VCl_3, $ZrCl_4$ etc.) and organometallic compounds ($AlEt_3$, $AlEt_2Cl$ etc.).

Example: Polyethylene, polystyrene, Polyvinyl chloride and polyacrylonitrile.

▶ Copolymer is formed by combining more than one monomer and the process is termed as copolymerization. When two or more monomers are combined to form copolymer is called bipolymer, terpolymer, quaterpolymer and so on periodically.

Example: Polyethylene-vinyl acetate (PEVA), nitrile rubber, nylon 6 and nylon 66.

2. Show that the rate of free radical polymerization process is dependent on the rate constants of these three steps (initiation, propagation and termination) and concentration of monomer and initiator.

Ans. In this free radical polymerization process monomer and catalyst (free radical generating molecule) were used. There are three primary steps i.e. initiation, propagation, and termination in free radical polymerization process. In the initiation step radicals are formed, in the propagation step molecular weight of living polymer increases and in termination step living polymer becomes dead and formed the polymer.

i) **Initiation:** It is the first step of radical polymerization. In this step an active centre is created from which a polymer chain is started to grow. Thermal or photochemical decomposition of organic peroxides, hydroperoxides, azo or diazo compounds, high-energy radiation, oxidation-reduction (redox) reactionsand electrochemical reactions are the ways for free radical creation.

initiator $(I) \rightarrow$ radical $(\overset{\bullet}{R})$ or radicals$(n\overset{\bullet}{R})$

radical $(\overset{\bullet}{R})$ + monomer $(M) \rightarrow$ monomer radical$(R\overset{\bullet}{M})$

Thus the rate equation is $R_i = \dfrac{d[R\overset{\bullet}{M}]}{dt} = nfk_i[I]$

For homolytic decomposition of initiator, $n = 2$

$$R_i = \frac{d[R\overset{\bullet}{M}]}{dt} = 2fk_i[I]$$

where k_i is rate constant and f is fraction of monomers take part in this step (generally the value of f is 0.3 to 0.8).

Actually, the initiation process proceeds through two sub-steps. In the first step initiator molecules create one or more than one radical depending on which type of catalyst is used. Then this radical will react with a monomer molecule giving a monomer radical. It was found that the formation of radical from initiator is the slowest step between these two steps.

ii) **Propagation:** It is the second step of radical polymerization. In this step the growth of a polymer chain occurs by successive addition of monomers.

$$R\overset{\bullet}{M} + M \rightarrow \overset{\bullet}{M}_2$$
$$\overset{\bullet}{M}_2 + M \rightarrow \overset{\bullet}{M}_3$$
$$\overset{\bullet}{M}_n + M \rightarrow \overset{\bullet}{M}_{(n+1)}$$

Thus the rate equation is

$R_p = \dfrac{-d[M]}{dt} = k_p[M][\overset{\bullet}{M}]$, where k_p is rate constant.

Radicals attack the loosely bound pi bond of the alkene forming another free radical and the process continuesuntil stop in the termination step.

iii) **Termination:** It is the final step of radical polymerization. In this step the growth of a polymer chain stops and obtains the final polymer. Combination of two living polymer chains, radical disproportionation and Combination of aliving polymer chain with an initiator radical or impurities or inhibitors will stop the polymerization process that is polymer chains terminated.

$$\overset{\bullet}{M}_x + \overset{\bullet}{M}_y \rightarrow M_{x+y}$$
$$\overset{\bullet}{M}_x + \overset{\bullet}{M}_y \rightarrow M_x + M_y$$
$$\overset{\bullet}{M}_x + \overset{\bullet}{R} \rightarrow M_xR$$

Thus the rate equation is

$$R_t = \frac{-d[\overset{\bullet}{M}]}{dt} = k_t[\overset{\bullet}{M}][x]$$

where k_t is rate constant and [x] is either another living polymer chain or initiator or inhibitor. When the reaction stop by other living polymer

chain, rate equation will be $R_t = \dfrac{-d[\dot{M}]}{dt} = 2k_t[\dot{M}]^2$

Using steady-state approximation,

Rate of initiation (R_i) = Rate of termination (R_t)

$$2fk_i[I] = 2k_t[\dot{M}]^2$$

$$[\dot{M}] = \sqrt{\dfrac{fk_i[I]}{k_t}}$$

Then the rate of polymerization

$$R_p = \dfrac{-d[\dot{M}]}{dt} = k_p[M][\dot{M}] = k_p[M]\sqrt{\dfrac{fk_i[I]}{k_t}}$$

$$R_p = k_p\left(\dfrac{fk_i}{k_t}\right)^{\frac{1}{2}} [M][I]^{\frac{1}{2}}$$

Thus, the rate of polymerization is depending on the rate constants of these three steps and concentration of monomer and initiator.

3. **When alternating copolymer and random copolymer will be form for**

   this equation $\dfrac{F_1}{1-F_1} = \dfrac{r_1\dfrac{[f_1]}{[1-f_1]}+1}{1+r_2\dfrac{[1-f_1]}{[f_1]}}$ **Where f_1 and f_2 are the mole fraction of**

   monomer in the reaction mixture and F_1 and F_2 are the mole fraction of

   monomer in thecopolymer formed and $r_1 = \dfrac{k_{11}}{k_{12}}$ and $r_2 = \dfrac{k_{22}}{k_{21}}$.

Ans. Alternating copolymer will form when $r_1 = r_2 = 0$ and random copolymer will form when $r_1 = r_2 = 1$.

4. **How many polymerizations techniques are there? Describe shortly with example.**

Ans. There are four polymerization techniques namely solution polymerization techniques, bulk polymerization techniques, emulsion polymerization techniques and suspension polymerization techniques.

▶ In solution polymerization a solvent is chosen in such a way that both monomer and initiator are dissolved in the solvent without reacting. Then the two solutions are mixed together to form the polymer.
   Example : Polyacrylonitrile, Polypropylene, Polystyrene, Polyisoprene, Polyethylene.

▶ In bulk polymerization process monomer is always in liquid state and the initiator is either in solid or in liquid state which is soluble in pure monomer. The reaction is initiated by thermally or photochemically. As there are no waste products or the final form of polymer does not need any purification, this process is environmentally friendly.

Example: Polyamide, Polycarbonate, Poly(methyl methacrylate), Nylon 6.

▶ In emulsion polymerization process monomer, emulsifier, initiator and dispersing medium are used. In particular cases modifiers are also used in the system. Water is normally used as a dispersing medium. The monomer is slightly soluble or insoluble in water but the initiator is soluble in water. Surfactant is used as an emulsifier which forms micelles around the monomer droplet. The molecular weight achieved by this method is higher than bulk or solution polymerization.

Example: Neoprene, Nitrile rubber, Polyvinyl fluoride, Polyvinyl acetate.

▶ In Suspension polymerization process monomer, stabilizing agent, initiator and dispersing medium are used. Mechanical agitation is required for this process. Water is normally used as a dispersing medium. The monomer is slightly soluble or insoluble in water and the initiator is soluble in monomer. Stabilizing agent is used to stabilize monomer-initiator droplets in the dispersion. The initiators used are benzoyl peroxide, AIBN, diacetyl peroxide, lauroyl peroxide and others. Stabilizers may be surfactants or other compounds. The polymers formed by this polymerization process have bead, pearl or granular like structures.

Example: Polyvinyl chloride, Polysulphide, Poly(styrene-acrylonitrile), Poly(methyl methacrylate).

Exercise

A **Multiple Choice Type Questions(MCQ) (Each Question carries 1 mark)**

1. Generally how many types of polymerization process are there?

    a. 2   **b.** 3

    c. 4   **c.** 8

2. What are the steps of polymerization?

    a. syndiotactic polymer and atactic polymer

    b. synthetic polymer and natural polymer

    c. initiation, propagation, and termination

    d. both b and c

3. What type of reaction is used in polymerization?

    a. chain reaction   **b.** ionic disproportionate reaction

    c. hydration reaction   **d.** hydrogenation reaction

4. How many types of addition polymerization are there?

    a. 6   **b.** 4

    c. 2   **d.** 3

5. For addition polymerization monomers must have
   a. radical
   b. different functional groups
   c. either a double bond or triple bond
   d. none of these

6. For condensation polymerization monomers must have
   a. different functional groups
   b. either a double bond or triple bond
   c. radical
   d. none of these

7. Free radical polymerization, cationic polymerization and anionic polymerization are the part of
   a. Addition polymerization
   b. Condensation polymerization
   c. Coordination polymerization
   d. Copolymerization

8. Radical is generated in the
   a. chain propagation step
   b. chain initiation step
   c. chain termination step
   d. none of these

9. BF_3 is used as catalyst in
   a. cationic polymerization
   b. anionic polymerization
   c. all of these
   d. none of these

10. *n*-butyl lithium is used as catalyst in
   a. anionic polymerization
   b. cationic polymerization
   c. all of these
   d. none of these

11. Condensation polymerization is also called
   a. radical polymerization
   b. ring closing polymerization
   c. step growth polymerization
   d. all of these

12. DNA is formed by
   a. step growth polymerization
   b. ring closing polymerization
   c. radical polymerization
   d. all of these

13. In condensation polymerization the number of monomer must be
   a. two or more
   b. one
   c. none of these
   d. all of these

14. Karl Ziegler and Giulio Natta were awarded the Nobel Prize in chemistry for the invention of
   a. radical polymerization
   b. step growth polymerization
   c. ring closing polymerization
   d. coordination polymerization

15. In copolymerization the number of monomer must be
   a. two or more
   b. one
   c. none of these
   d. all of these

16. Generally how many polymerization techniques are there
   a. 4
   b. 6
   c. 8
   d. 10

17. Generally in solution polymerization a solvent is chosen in such a way that both monomer and initiator are dissolved in the solvent

  a. with reacting

  b. without reacting

  c. monomer and initiator are not dissolved

  d. none of these

18. In bulk polymerization process monomer is always in liquid state and the initiator is either in solid or in liquid state which is soluble in pure monomer

  a. true   **b.** false

  c. all of these   **d.** none of these

19. In which polymerization process surfactant is used?

  a. Emulsion polymerization techniques **b.** Solution polymerization techniques

  c. Bulk polymerization techniques   **d.** Suspension polymerization techniques

20. Poly(methyl methacrylate)(PMMA) is usually obtained by the

  a. addition polymerization   **b.** condensation polymerization

  c. coordination polymerization   **d.** copolymerization

B **Short Answer Type Questions (Each question carry either 2 or 3 marks)**
Write short notes on the following

  1. Addition polymerization

  2. Condensation polymerization

  3. Coordination polymerization

  4. Copolymerization

  5. Solution polymerization techniques

  6. Bulk polymerization techniques

  7. Emulsion polymerization techniques

  8. Suspension polymerization techniques

C **Long Answer Type Questions (Each question carry 5 marks)**

  1. What do you understand by free radical polymerization? Explain the mechanism with example.

  2. What do you understand by cationic polymerization? Explain the mechanism with example.

  3. What do you understand by Anionic polymerization? Explain the mechanism with example.

  4. Describe the types of polymerization process.

  5. What do you understand by addition polymerization? Explain the mechanism with example.

6. What do you understand by condensation polymerization? Explain the mechanism with example.

7. What do you understand by coordination polymerization? Explain the mechanism with example.

8. What do you understand by copolymerization? Explain the mechanism with example.

9. Describe the polymerizations techniques.

Answers

A

1. a	2. c	3. a	4. d	5. c	6. a	7. a	8. b	9. a	10. a
11. c	12. a	13. a	14. d	15. a	16. a	17. b	18. a	19. a	20. a

4

Crystallization and Crystallinity

Crystallization is the process by which crystals are formed and degree of crystallinity is a value to measure the amount of crystals formed. Polymers are either crystalline or amorphous. Polymer structure and intermolecular forces are the two important parameters for the origination of crystallinity into the polymer. According to Bragg's diffraction law ($n\lambda = 2d\sin\theta$ where n is an integer, λ is the wavelength of the X-rays radiation and d is the spacing between two successive layers), a material which has layer or ordered structure is crystalline. But for polymers the crystallinity is due to the presence of **lamella**. In this regard it should be kept in mind that there is no any polymer which is 100% crystalline. If you see the molecular structure of a polymer you can easily understand it. In a polymer you will find random polymer chain arrangement and ordered polymer chain arrangement. The amorphous nature is due to random polymer chains arrangement and crystalline nature is due to ordered polymer chains arrangement in the polymer. When a polymer is formed, few random polymer chain or parts of polymer chains are always present there. This restricts the polymer for 100% crystallisation.

Example :

- **Crystalline Polymer :** High density polyethylene (HDPE), polypropylene, syndiotactic polystyrene, nylon, polybutylene terephthalate (PBT), kevlar, Nomex.
- **Amorphous Polymer :** Polyvinyl chloride (PVC), poly (methyl methacrylate), atactic polystyrene, polycarbonate, polybutadiene, polyisoprene.

Pic 4.1 Crystalline structure formed by single polymer chain

Pic 4.2 Amorphous polymer structure

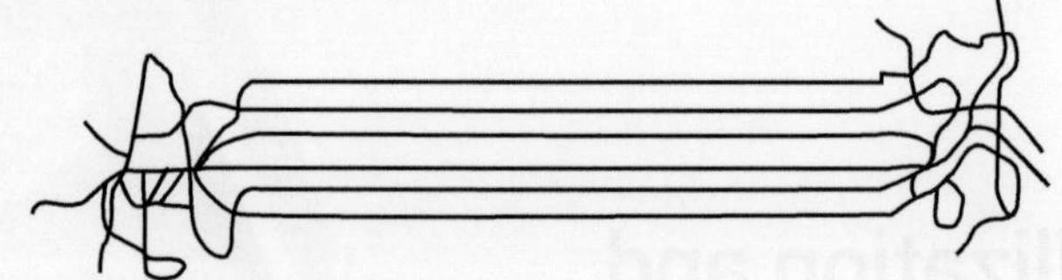

Pic 4.3 Crystalline structure formed by many polymer chains

In crystalline polymers lamellae thickness vary from 10 to 20 nm.

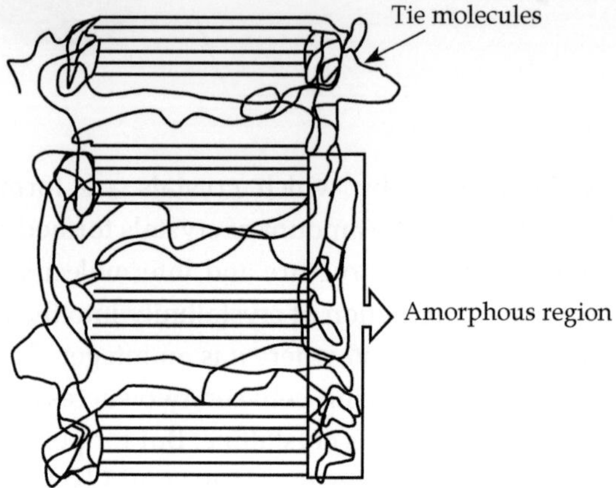

Pic 4.4 Crystalline structure formed
by interlamellar ties of polymer chains

 # Tie molecules

It is the polymer chain taking part in both polymer amorphous and crystalline regions formation and should pass through the amorphous region to the crystalline region.

The origin of crystallinity in nylon 66 is due to intermolecular forces. Polar amide groups in the nylon 66 backbone chain form strong hydrogen bonding and raise crystallinity.

The oxygen atom of carbonyl group forms hydrogen bond with the hydrogen atom of amide group in nylon 66.

Check the molecular structures of Polypropylene, syndiotactic polystyrene, nylon, kevlar, nomex, Poly(methyl methacrylate), Atactic polystyrene, Polycarbonate, Polyisoprene and polybutadiene and find out which one is crystalline and which one is amorphous.

Now we understand that polymer contains both amorphous region and crystalline region. There are few instrumental techniques by which we can measure the crystallinity of the polymer. Differential scanning calorimeter (DSC) measures the crystalline melting point as well as degree of Crystallinity of polymer. But there are other instrumental techniques namely X-ray diffraction (XRD), nuclear magnetic resonance (NMR) spectroscopy, Raman spectroscopy and neutron scattering and Fourier transform infrared (FTIR) spectroscopy by which crystallinity in polymer is measured. A quick judge whether the polymer is crystalline or not can be done using transmission electron microscope (TEM). TEM gives the idea extremely in short range not in bulk.

DSC measures the amount of heat change when a polymer melted or crystallised and a reference of 100% crystalline material will compare to calculate the degree of crystallinity of the experimental unknown polymer. DSC measures the crystallinity of the melted part which does not include the contributions of isolated ordered chains to short-range order of polymer. The oxygen atom of carbonyl group forms hydrogen bond with the hydrogen atom of amide group in nylon 6,6. Check the molecular structures of Polypropylene, syndiotactic polystyrene, nylon, kevlar, nomex, Poly(methyl methacrylate), Atactic polystyrene, Polycarbonate, Polyisoprene and polybutadiene and find out which one is crystalline and which one is amorphous.

XRD measures long-range order or intermolecular order in the polymer chains. By calculating the area under the amorphous and crystalline region, the degree of crystallinity of the experimental unknown polymer is calculated.

Actually DSC and XRD are the two main instruments to measure the polymer crystallinity. The other instrumental techniques are helpful to give more information. Solid-state NMR is generally used to calculate the crystal packing environments of polyamides. These spectra are complicated and have wide limitation. Raman spectroscopy is used for crystalline polymers where an energy exchange takes place between the incident photons and the vibrational energy levels of the polymer. Strong Raman scattering is observed for polymer having nonpolar molecular groups. Neutron scattering gives information of a polymer on its chain correlations over a wide range of distances. FTIR is also useful for analysing crystalline polymers of short-range order.

As the basic principle for the measurement of polymer crystallinity by using different instruments are different therefore the method used to measure crystallinity needs to be specified for the meaningful reported results.

4.1 Measurement of Crystalline Melting Point and Degree of Crystallinity

The degree of crystallinity sometimes refer percentage of crystallinity of a polymer is a relative quantity. It is estimated by different methods and have a value ranging from 10% to 80% generally.

Heat content and specific volume are the two thermodynamic properties which vary with crystallization. So, the measurement of specific volume or heat flow, measures the degree of crystallinity in the polymer. Here mainly three methods will be discussed for degree of crystallinity measurement and one method for crystalline melting point measurement.

(A) (i) **Measurement of the degree of crystallinity in polymers from density measurements :** A typical specific volume vs. temperature plot of different states of material is presented below. Specific volume of liquid and crystal increases linearly with increasing temperature having different slopes. But for a mixture of liquid + crystal system the increment is like a curve.

T_g, T_m and T_x are the glass transformation temperature, melting temperature and the temperature where the specific volume of the crystals and liquid are same. The point X is experimentally very difficult to attain for complicated structure polymers but is possible for simple linear polymers. Let X = volume fraction of crystals (degree of crystallinity) in mixture with liquid or glass, V = specific volume of mixture, V_{lg} = specific volume of liquid or glass, whichever is appropriate to the given temperature, V_s = specific volume of crystals. Then the degree of crystallinity will be $X = (V_{lg} - V)/(V_{lg} - V_s)$. This method is applicable for polyethylene.

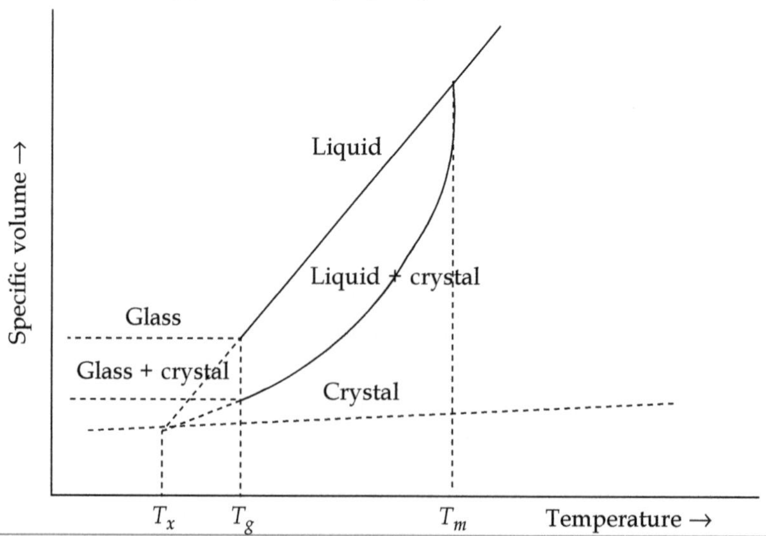

(ii) **Measurement of the degree of crystallinity in polymers from XRD patterns:** The following figure is for XRD patterns of amorphous, semi-crystalline and crystalline polymer. For understanding you can think sharp peak is originated from crystalline region and broad peak is originated from amorphous region. Integrating the peak area (peak intensity may also use) of amorphous and crystalline region degree of crystallinity in polymers is calculated.

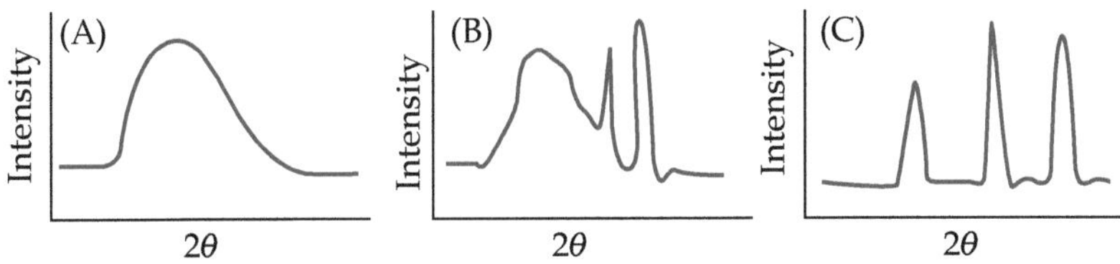

Pic 4.5 XRD patterns of (A) amorphous (B) semi-crystalline (C) crystalline polymers.

Here a very simple XRD pattern is presented for stress-free understanding the calculation of degree of crystallinity in polymers. You may or may not get this XRD pattern of any experimental polymer. Let c and a are the integrated peak area of crystalline and amorphous region respectively, then degree of crystallinity of this polymer will be $\dfrac{c}{c+a} \times 100\%$. For this type of XRD pattern you can easily calculate the integrated peak area by pointing it in a graph paper and counting each small division. But in reality XRD pattern are not so simple and the software will do this job for you.

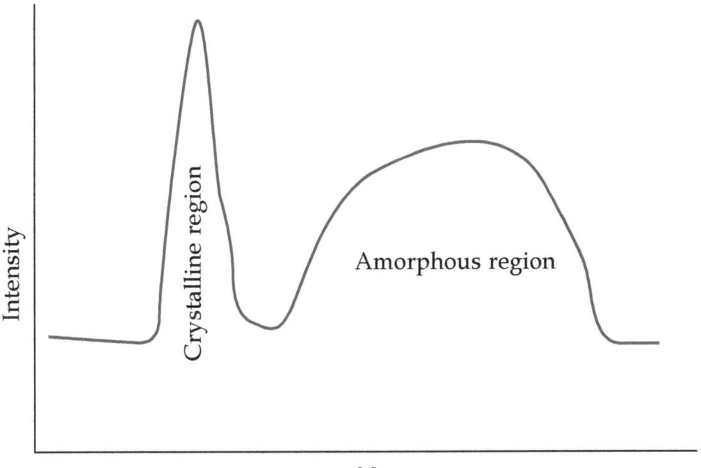

(B) **Measurement of the crystalline melting point and degree of crystallinity in polymers from the DSC thermograms:** The DSC thermogram of polymer is presented below. It is very clear that the crystalline melting point of polymer is directly measured from the melting peak position of the DSC thermogram. But for the degree of crystallinity you have to do a little bit of mathematics.

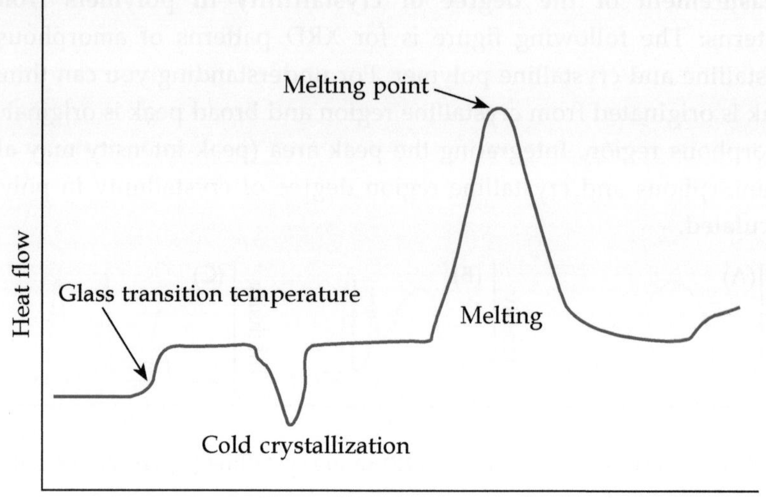

If Q_m, Qc and Q_m^0 are the heat of melting, heat of cold crystallization and the heat of melting of the polymer when it is 100% crystalline (reference). Then the degree of crystallinity of the polymer will be $\dfrac{Q_m - Q_c}{Q_m^0} \times 100\%$. Remember that cold crystallization peak may or may not be observed always. The reference heat of melting (Q_m^0 in J/g) for nylon 66, polyethylene, polypropylene and polyethylene terephthalate are 255.8, 293.6, 207.1 and 140.1 J/g respectively.

One of the method for polymer crystallization is by the cooling of polymer melt. During cooling polymer melt changes from a liquid/viscous liquid phase to a ordered solid phase. The temperature at which this transition occur is called the **crystallization melting point of that polymer.**

■ **Melting points of polymers :**

Polymer name	Melting point (T_m)
Polyethylene	141°C
PVC	220°C and 305°C
isotactic polystyrene	242°C
isotactic polypropylene	180°C
Nylon 6,6	272°C
PET	260°C
PEG	60°C
Polycaprolactum	218°C
PVA	230°C and 270°C
Poly(vinylidene fluoride)	176°C

Remember that polymer melting peak is a bread peak.

4.2 Morphology of Crystalline Polymers

Morphology of crystalline polymers mainly depends on molecular configuration of repeat units, geometrical arrangements of neighbouring groups in molecule and molecular packing in unit cell that is chain length, chain branching and interchain bonding. Crystallinity originates either from cooling the melted polymer or nucleation in solution.

Spherulite forms when lamella of polymer chains forms a sphere like morphology (in 2D it looks like a bicycle ring with spokes.

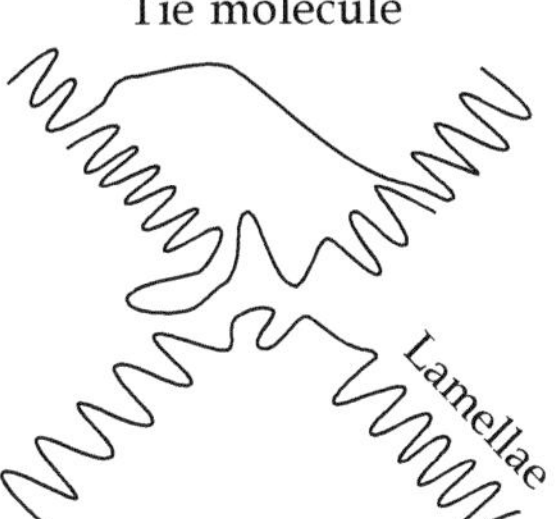

Pic 4.6 Polymer spherulite

[spherulites have a diameter ranging from 1μm to 10μm]

In the lamella if the polymer chains go out and come back consecutively or with a gap to the lamella is called Flory switchboard model. Polymer morphology is seen in SEM (scanning electron microscope), TEM (transmission electron microscope) and AFM (atomic force microscope). A SEM image of crystalline polymer is shown below.

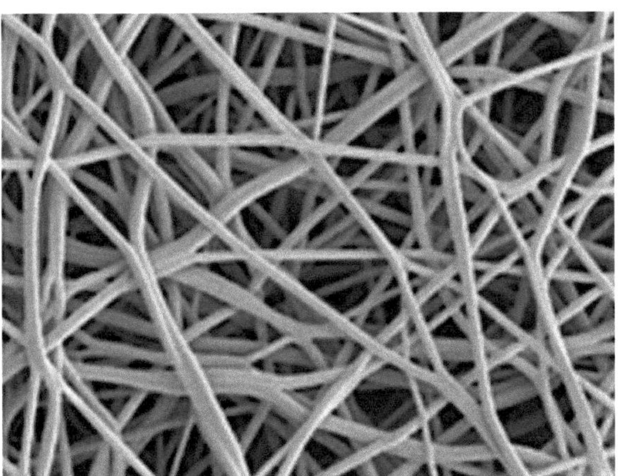

Pic 4.7 Crystalline polymer

4.3 Factors Affecting Crystalline Melting Point

Crystalline melting point of a polymer is the temperature at which crystals melt. Thus it is related to the nature of crystal that is the nature of formation and properties of crystals. Degree of crystallinity increases when crystals are prepared slowly. It will affect the crystal melting point. Beside that impurity, fillers (generally clay, carbon nanotube, metal nanoparticles are used as filler) and processing techniques affect the crystalline melting point of the polymer. It also depends on strain-induced crystallization, size and structure of the crystal, nucleating agent, tacticity of the polymer, bulky groups in the polymer chain and the length of side chains.

1. What is the procedure of increasing crystallinity of polymer samples?

Ans. No polymer is 100% crystalline. It has both amorphous part and crystalline part. Crystalline melting point of a polymer is the temperature at which crystals melt. Thus it is related to the nature of crystal that is present in polymer. The nature of crystal mainly depends on its formation and properties of crystals. Degree of crystallinity increases when crystals are prepared slowly. It will increase the crystal melting point. Beside that impurity, fillers (generally clay, carbon nanotube, metal nanoparticles are used as filler) and processing techniques affect the crystalline melting point of the polymer.

2. How does XRD pattern of amorphous and crystalline polymers look like?

Ans. XRD pattern of amorphous and crystalline polymers are shown below

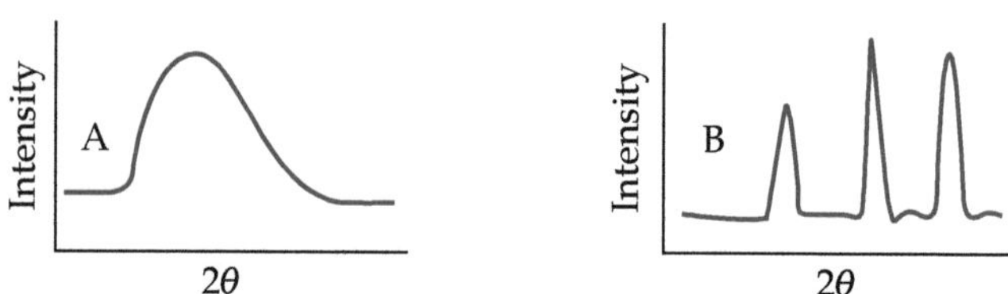

XRD patterns of (A) amorphous (B) crystalline polymers

In amorphous polymer a broad peak is observed and in crystalline polymer sharp peak/peaks are observed. The sharp peak/peaks are come from lamellar structure of the polymer chain/chains.

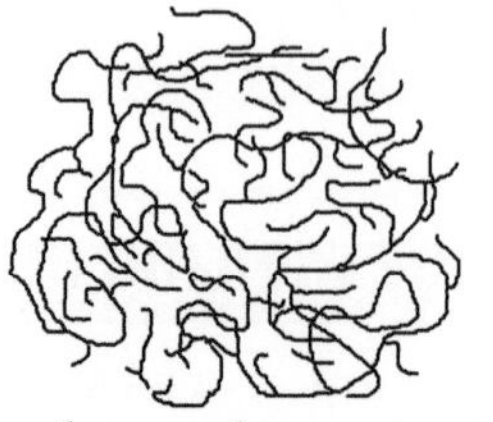

Amorphous polymer structure   Crystalline structure formed by single polymer chain

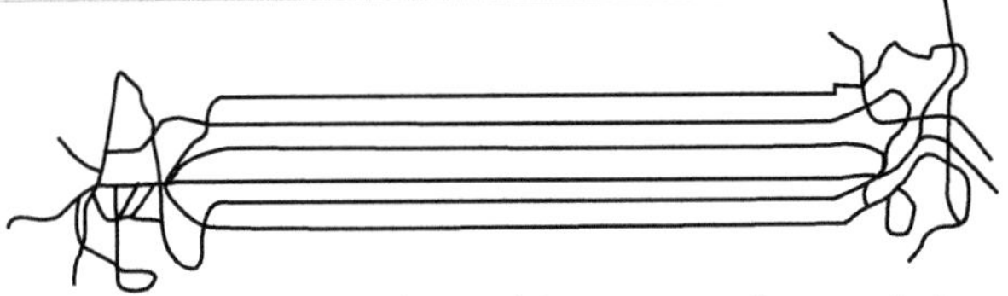

Crystalline structure formed by many polymer chains

3. How will you calculate crystalline melting temperature from DSC curves?

Ans. A typical DSC thermogram of polymer is presented below. It is very clear that the crystalline melting point of polymer is directly measured from the melting peak position of the DSC thermogram.

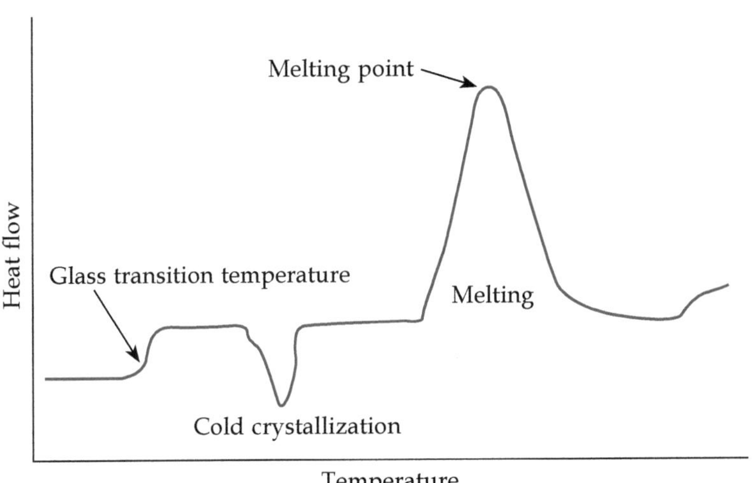

If Q_m, Q_c and Q_m^0 are the heat of melting, heat of cold crystallization and the heat of melting of the polymer when it is 100% crystalline (reference). Then the degree of crystallinity of the polymer will be $\dfrac{Q_m - Q_c}{Q_m^0} \times 100\%$. Cold crystallization peak may or may not be observed always. The reference heat of melting (Q_m^0 in J/g) for nylon 6,6, polyethylene, polypropylene and Polyethylene terephthalate are 255.8, 293.6, 207.1 and 140.1 J/g respectively. Using these data crystalline melting temperature of polymer from DSC curves will be calculated.

4. What is Polymer spherulite?

Ans. Crystallinity originates either from cooling the melted polymer or nucleation in solution. Spherulite forms when lamella of polymer chains forms a sphere like morphology (in 2D it looks like a bicycle ring with spokes.

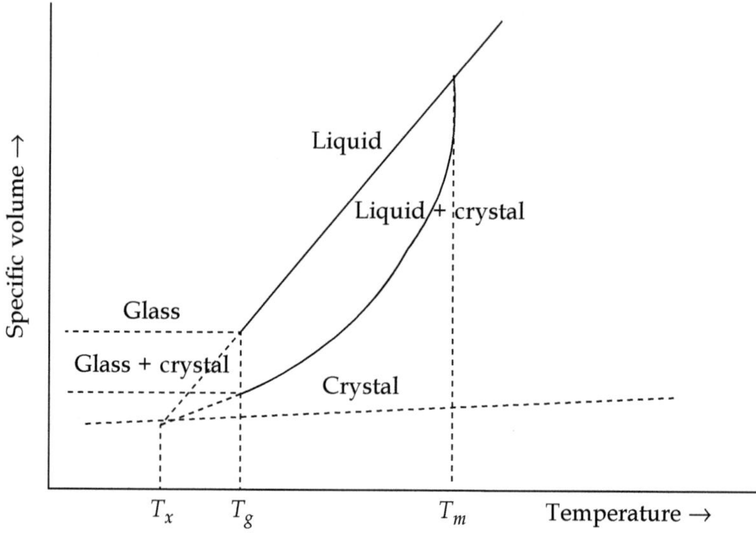

5. How will you measure the degree of crystallinity in polymers from density measurements?

Ans. A typical specific volume vs. temperature plot of different states of material is presented below

Specific volume of liquid and crystal increases linearly with increasing temperature having different slopes. But for a mixture of liquid + crystal system the increment is like a curve. T_g, T_m and T_x are the glass transformation temperature, melting temperature and the temperature where the specific volume of the crystals and liquid are same. The point X is experimentally very difficult to attain for complicated structure polymers but is possible for simple linear polymers. Let X = volume fraction of crystals (degree of crystallinity) in mixture with liquid or glass, V = specific volume of mixture, V_{lg}= specific volume of liquid or glass, whichever is appropriate at the given temperature, V_s = specific volume of crystals. Then the degree of crystallinity will be $X = (V_{lg} - V)/(V_{lg} - V_s)$.

Exercise

A Multiple Choice Type Questions(MCQ) (Each Question carries 1 mark)

1. Crystallinity defines the degree of
   a. long-range order in a material
   b. short-range order in a material
   c. intramolecular hydrogen bonding in a material
   d. intermolecular hydrogen bonding in a material

2. Crystallinity of polymers measured by
   a. C, H, N analysis
   b. XRD spectra
   c. centrifuging
   d. none of these

3. According to Bragg's diffraction law $n\lambda = 2d\sin\lambda$ (where n is an integer, λ is the wavelength of the x-rays radiation and d is the spacing between two successive layer)
   a. true
   b. false
   c. $n\lambda = d\sin\lambda$
   d. both b and c

4. The origin of Crystallinity in nylon 6, 6 is due to
   a. H atom
   b. intramolecular forces
   c. N atom
   d. intermolecular forces

5. If there is broad peak in XRD spectrum then the polymer is
   a. amorphous
   b. crystalline
   c. semi crystalline
   d. all of these

6. TEM gives the idea about crystallinity of polymers in
   a. long range
   b. extremely short range not in bulk
   c. only bulk
   d. long range and bulk

7. Crystalline melting point is affected mainly by
   a. fillers
   b. monomer
   c. initiator
   d. none of these

8. Does the crystallinity of polymer depend on sample thickness (consider the polymer structure is homogeneous)
   a. no
   b. yes
   c. to some extent
   d. none of these

9. Spherulite is formed when lamella of polymer chain forms
   a. linear morphology
   b. sphere like morphology
   c. dumb bell shape morphology
   d. none of these

10. Can any natural polymers be 100% crystalline
    a. no
    b. yes
    c. neoprene
    d. Styrene-butadiene rubber

B Short Answer Type Questions (Each question carry either 2 or 3 marks)

1. What is the procedure of increasing crystallinity of polymer samples?
2. How you will measure the degree of crystallinity in polymers from density measurements?
3. Can plastic deformation in polymer increase crosslinking density of the polymer?
4. Can the crystallinity of a polymer be improved or restored from its melt?
5. How can shear rate and temperature affect polymer nucleation during crystallization?
6. Does polymer crystallinity improve the electrical conductivity of it?
7. What is the effect of size and morphology of polymer spherulites on the elongation and tensile strength of membranes?
8. What is cold crystallization?
9. Why one polymer shows the cold crystallization peak and the other one doesn't?
10. Show schematically the polymer chains to originate crystallinity.
11. What are the methods to measure crystallinity of polymers?
12. How does XRD pattern of crystalline polymers look like?
13. How does DSC pattern of crystalline polymers look like?
14. What is Polymer spherulite?

C Long Answer Type Questions (Each question carry 5 marks)

1. What are the factors that affect the crystalline melting point?
2. Explain morphology of crystalline polymers with drawing picture.
3. Hoe will you calculate crystallinity of a polymer from XRD pattern?
4. Hoe will you calculate crystallinity of a polymer from DSC curve?
5. Draw the XRD patterns of (A) amorphous (B) semi-crystalline (C) crystalline polymer.
6. How to calculate glass transition temperature from DSC curves?

Answers

1. a	2. b	3. a	4. d	5. a	6. b	7. a	8. a	9. b	10. a

5

Nature and Structure of Polymers

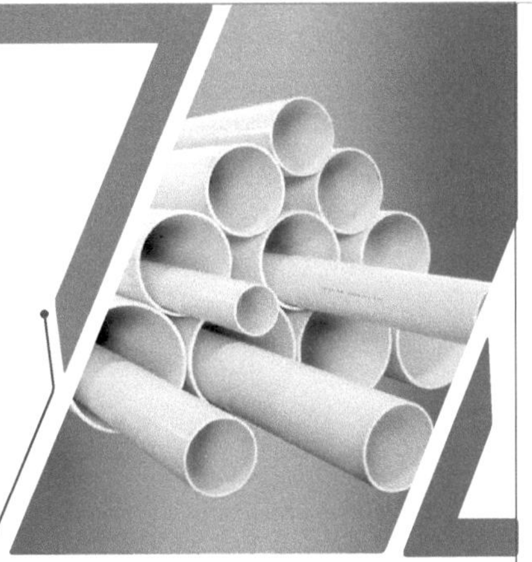

Actually polymers' nature will depend on the polymer's structure. Polymer may be crystalline or amorphous, linear or nonlinear, conducting or non-conducting, hydrophilic or hydrophobic, elastic or viscous; everything will depend on its structure and bonding. Polymers are formed by combining many numbers of monomer units and the structure of the monomer unit directs the polymer to linear, branch, cross-linked and network structure. The structure of small molecules can be predicted accurately but for polymer molecules, it is highly unlikely to predict its structure so accurately. Beside that the small molecules have definite molecular weight. In the case of polymers, the molecular weight is not definite, it is a distribution. It means when polymers are formed from monomers, if in the polymer chains the number of monomer units say 100, 120, 150, 200 etc. then the molecular weight of polymer chains will be equal to the molecular weight of monomer multiplied by the number of above monomer units respectively. This indicates that there are a range of chain length in polymer and the arrangement of these polymer chains individually or all together have importance in polymer structure and hence the nature of the polymers.

Let's start thinking from the structure of small molecules. Suppose in same molecular formula the material has three states solid, liquid and gas. Take an example H_2O with its three states namely ice, water and water vapour. Depending on the packing (arrangement due to molecular force) of H_2O molecule it is either solid or liquid or vapour *i.e.*, structure direct nature and properties of the materials.

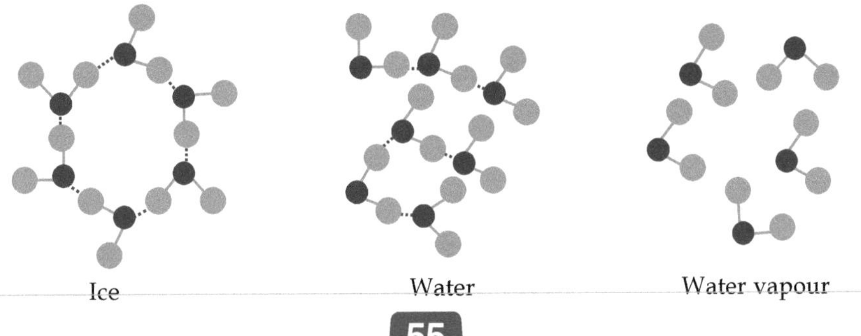

Ice   Water   Water vapour

■ Now come to polymer. Polymers are not like small molecules in terms of molecular weight. Practically the molecular weight of a polymer is not a single value number, it is a distribution. In the following figure you can see that there are 6 polymer chains of different numbers of monomer units. The monomer unit of the polymer gradually increases which increases the chain lengths and molecular weight of the polymer. It is worthy to mention here that the conductivity of conducting polymer increases with polymer chain lengths due to extended π—π conjugation. Take an example for conception to understand the polymer chain length. Let assume a monomer unit length (keeping in mind that size is three dimensional and length is one dimensional) is 134 pm (a) and distance (bond length) between two monomer units is 154 pm (*d*). In a polymer chain if 1000 monomer units (*n*) are present then polymer chain length will be equal to $[(a \times n) + (n - 1) \times d]$ *i.e.*, 286 nm. Now you get little idea how long a polymer chain can be. Again the number of monomer units in the polymer chain is variable. If there are few monomer units in the composed chain, it is called an oligomer not a polymer. In this respect dimers, trimers and tetramers are oligomers. There are no particular numbers for which a composed chain will be called an oligomer. Sometimes eight monomer units are composed to form a molecule is also called oligomer. The properties of oligomer and polymer are different. For conducting polymer, oligomer is non-conducting but polymer is conducting.

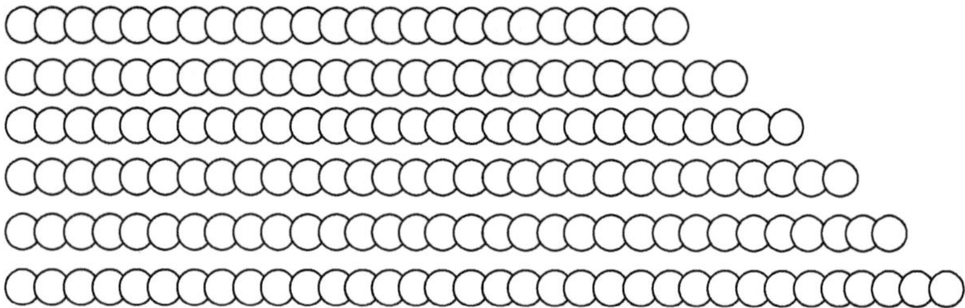

Polymer may remain in solid state or in solution (dissolve in solvent). There are many methods for polymer synthesis namely polymer synthesis in homogeneous medium, polymer synthesis in heterogeneous medium, solid state polymer synthesis, polymer synthesis by interfacial method and so on. Polymers are also processed in a number of ways. Generally in solution polymer chains may remain expanded (in good solvent) or collapsed, precipitated (in poor solvent).

Good solvent

Poor solvent

Polymer may be hydrophobic or hydrophilic and thus a solvent will be good or poor totally depending on the polymer solvent interaction. If a solvent is good for one polymer may be it will be bad for another polymer.

The most commonly used solvents for polymer processing are tetrahydrofuran (THF), dimethylacetamide (DMAc), N-methyl-2-pyrrolidone (NMP), dimethyl formamide (DMF), Dimethylsulfoxide (DMSO), 1,2,4-trichloro benzene (TCB), Triethylamine (TEA), dichloromethane(DCM).

See the following table for the processing solvent of definite polymer.

Polymer	Solvent	Polymer	Solvent
Poly(acrylates)	DMAc, DMF	Polyethylene glycol (PEG)	H_2O, DMF
Polyacrylonitrile (PAN)	DMF	Poly isoprene	THF, Toluene
Poly(vinyl chloride) (PVC)	THF	Polylactic acid (PLLA)	THF
PVDF	DMSO	Polysaccarides	H_2O, DMSO
Polyacrylamide (PAM)	H_2O	Polystyrene	THF
Polyaniline	NMP, *m*-cresol	Polysulfone	DMF
Polybutadiene	THF	Polythiophene	TCB, $CHCl_3$
Polycarbonate	DCM, THF	Polyurethane	THF
Poly(dimethyl siloxane) (PDMS)	Toluene	PVAc	THF
Polyester	THF	PVA	H_2O
Polyethylene	TCB	Polyvinylpyrrolidone (PVP)	DMAc, H_2O

When solvent is evaporated, the polymer will become solid. A single polymer molecule may have varieties of structure linear, branched, cyclic, network, comb, brush, dumbbell, H-shape and so on.

i. Linear polymer :

Example : High density polyethylene, polystyrene poly(vinyl chloride), Nylon

ii. Branched polymer :

Example : Low density polyethylene, copolymer of ethylene and 1-butene

iii. Cyclic polymer :

Example : *Cyclo*-poly (tetrahydro furan), *cyclo*-poly(styrene), *cyclo*-poly (propylene glycol).

iv. Network polymer :

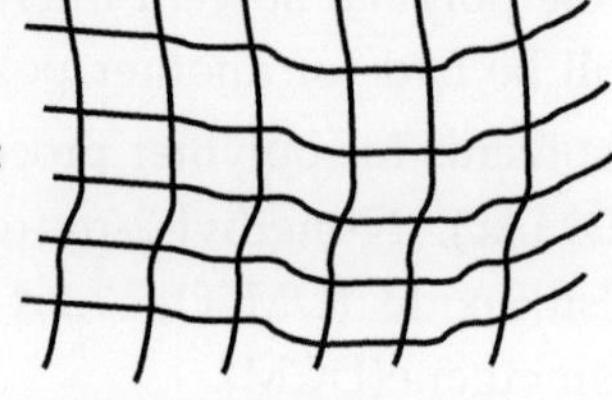

Example : Vulcanized rubber, bakelite, melamine, formaldehyde resins.

v. Comb polymer :

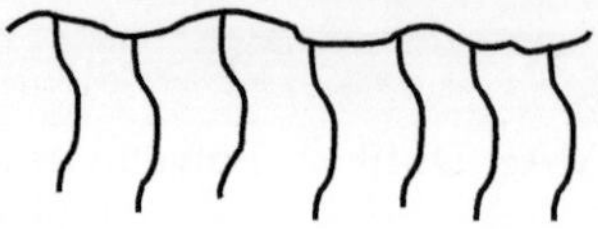

Example : Poly(L-lactide) on poly (2-hydroxyethyl methacrylate), poly (L-lactic acid) on poly (vinylidene fluoride)

vi. Brush polymer :

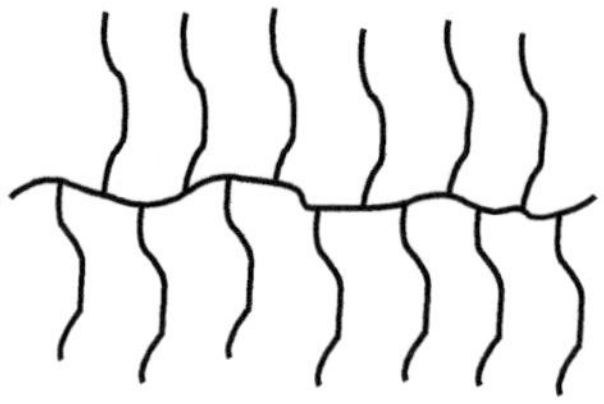

Example : Poly (2-hydroxyethyl methacrylate), poly[N-(2-hydroxy propyl) methacrylamide], poly (carboxybetaine acrylamide).

vii. H-shaped polymer :

Example : Polystyrene on poly(ethylene glycol);

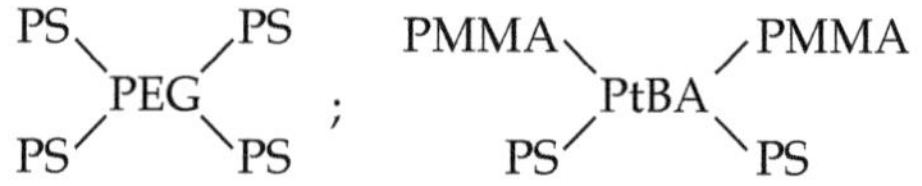

viii. Dumbbell polymer :

Example : Polystyrene on linear polystyrene (core) with polystyrene-co-trimethoxysilyl propylacrylate (shell).

5.1 Piezoelectric Polymer

Piezoelectricity was discovered in 1880 by two french scientists Jacques Curie and Pierre Curie. When mechanical pressure is applied to piezoelectric polymer, it generates electric potential across certain faces of the polymer. Mechanical stress applied to piezoelectric polymer generates electricity.

Example : poly(vinylidene fluoride), poly(vinylidene fluoride-*co*-tri fluoroethylene), poly (L-lactic acid), nylon-11.

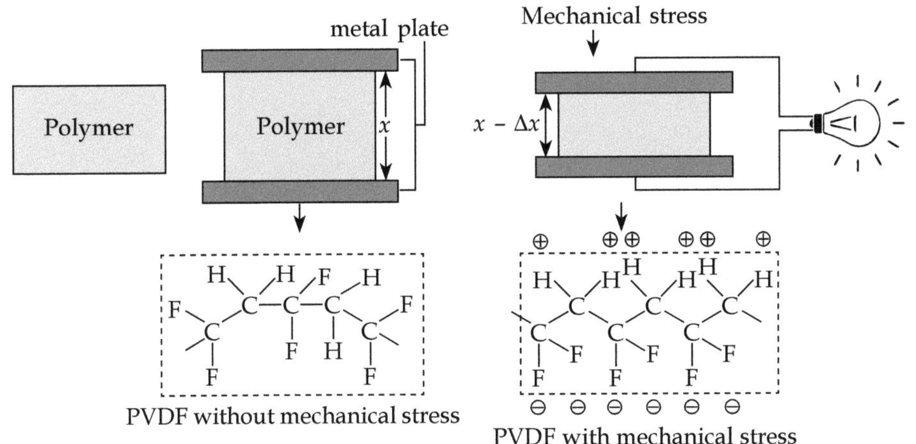

Piezoelectric polymers are widely used in biosignal acquisition, tissue regeneration and many other biological applications.

5.2 Optically Active Polymers

Optically active polymers are generally prepared by polymerization of optically active monomers.

Example : Polyacetylene, polyurethane, polyaniline poly(amide-imide)

application : Enantioselective separation, asymmetric catalysis, chemical and biological sensing.

1. What is the difference between polymer and macromolecule?

Ans. Polymer is a macromolecule which is composed of a large number of repeating monomer units. In this respect one can say polymer is macromolecule but macromolecule may or may not be polymer. Macromolecule is also a large molecule having diameter 10-1000 nm. Polymer is always made from monomers but macromolecules may or may not be made from monomers.

Examples of polymers are polyethylene, polyaniline and polypyrrole whereas Proteins, DNA, RNA, and plastics are the examplesof macromolecules.

2. **Explain good solvent and poor solvent. Which are good solvent and poor solvent for polyaniline?**

Ans. Generally, in a solution polymer chains remain as expanded or collapsed/precipitated.

Good solvent

Poor solvent

When the polymer chains remain expanded in a solvent, the solvent is called good solvent for that polymer. Again, when the polymer chains remain collapsed/precipitated in a solvent, the solvent is called poor solvent for that polymer.

Good solvent for polyaniline: formic acid, NMP, m-cresol.

Poor solvent for polyaniline: ethanol, acetone, water.

3. **Explain structure dependence conductivity of polymers.**

Ans. When there are labile electrons (generally π electrons) in polymer chains, polymers become conducting. Polyacetylene, polyaniline, polypyrrole and polythiophene are the examples of conducting polymers.

Now consider any one conducting polymer and its derivative. When there are substitution/s in the conducting polymer, the electron density in the polymer chains vary (either increase or decrease depending on the nature of the substitution groups). When the density of labile electrons in the polymer chains increases, conductivity of the polymer increases.

4. **What are linear polymers? How does it differ in property with branched polymers?**

Ans. In linear polymers the repeating units are joined together in such a way (end to end)that it forms a single flexible chain.

Linear polymer :

polyethylene, PVC, polystyrene, and polyamidesare the examples of linear polymers.

In branched polymers, branches are present which make more free space in the polymer resulting in a loosely packed polymer. It has lower density, melting point, boiling point and crystallinity than linear polymer. Starch, glycogen are examples of branched polymers.

Branched polymer :

Exercise

A Multiple Choice Type Questions(MCQ) (Each Question carries 1 mark)

1. Carbohydrates, Lipids, proteins and nucleic acids are
 a. macromolecules
 b. polymers
 c. small molecules
 d. all of these

2. When all the subunits are of the same type the macromolecules are called
 a. fragments
 b. ionomers
 c. radicals
 d. polymers

3. If there is conjugation in the polymer, it is called
 a. conducting polymer
 b. non-conducting polymer
 c. insulator
 d. both b and c

4. Macromolecules have a diameter ranging from
 a. 10^{-3} to 10^{-2} mm
 b. 10^{-5} to 10^{-3} mm
 c. 10^{-2} to 10^{-1} mm
 d. 10^{1} to 10^{2} mm

5. The basic polymer structures are
 a. branched, crosslinked, and networked
 b. linear, branched and crosslinked.
 c. linear, branched, crosslinked, and networked
 d. linear, crosslinked, and networked.

6. Molecular weight of polymer is
 a. a distribution
 b. a single valued number
 c. fraction
 d. none of these

7. Which can form cyclic polymer
 a. polypyrrole
 b. polyesters
 c. polythiophene
 d. pvc

8. Which can form star polymer
 a. poly (ethylene oxide)
 b. polyvinyl alcohol
 c. natural rubber
 d. natural silk

B Short Answer Type Questions (Each question carry either 2 or 3 marks)
 Write short notes on the following

1. What is the difference between polymer and macromolecule?

2. Good solvent and poor solvent.
3. Draw the different structures of polymer molecules.
4. Structure dependence conductivity of polymers.
5. Structure dependence crystallinity of polymers.
6. Structure dependence glass transition temperature of polymers.
7. Structure dependence morphology of polymers.
8. Structure dependence mechanical properties of polymers.
9. Structure dependence thermal stability of polymers.

C Long Answer Type Questions (Each question carry 5 marks)

1. Describe the nature and structure of polymers.
2. Draw the different structures of polymer molecules with examples.

Answers

A

1. a	2. d	3. a	4. b	5. c	6. a	7. b	8. a

6

Determination of Molecular Weight of Polymers

Small molecules like sodium chloride whenever and however is prepared one molecule of sodium combined with one molecule of chlorine forming a definite molecular weight of sodium chloride. But polymers have distribution in polymeric chain lengths. In one polymeric chain the number of monomer units differs from other polymeric chain. This is why the length of polymer chains are different with each other. The changes in polymer chain length changes the properties of polymers. The variation of chain lengths originates in the distribution in Molecular weights of polymers. There are many terms to express the molecular weight of polymers namely **number average molecular weight, weight average molecular weight, Z average molecular weight and viscosity average molecular weight**.

Take this story to understand the molecular weight of polymers. Suppose there are deers, elephants, tigers and lions in a jungle. The jungle is square and each animal lives particularly in one arm of the square. Thus the jungle looks like a deer jungle from one arm of the square, elephants jungle from another arm and so on. Though the jungle contains four different animals but it apparently behaves like an individual animal jungle. Due to seasonal or any other effect the animals will interchange their position. Polymers also behave like this. Any types of perturbation (shear, tensile and others forces) will impact the polymer chains behave like animals. Polymers have smaller chains, bigger chains and intermediate chains into itself. Few properties of the polymers are related to smaller chains, another few properties are related to bigger chains and so on. For example weight average molecular weight is useful for the expected statistical size of the polymer, whereas the number average molecular weight is for the chain length. The weight average is suitable for higher molecular weight molecules whereas the number average is suitable for lower molecular weight molecules. This is why different types of molecular weight of polymers are used.

It will be easy to learn molecular weight distribution of the polymers when studied with examples. It is previously discussed that polymers have varieties of polymer chains and the number (N_i) of such chains are also varied. In the

"

following table the weight of polymer chain and the number of such chains are presented

Number of chains (N_i)	Molecular weight of each chains (M_i)	Total mass (N_iM_i)
(N_1) = 1	(M_1) = 20000	(N_1M_1) = 20000
(N_2) = 4	(M_2) = 40000	(N_2M_2) = 160000
(N_3) = 8	(M_3) = 60000	(N_3M_3) = 480000
(N_4) = 16	(M_4) = 80000	(N_4M_4) = 1280000
(N_5) = 20	(M_5) = 100000	(N_5M_5) = 2000000
(N_6) = 16	(M_6) = 120000	(N_6M_6) = 1920000
(N_7) = 8	(M_7) = 140000	(N_7M_7) = 1120000
(N_8) = 4	(M_8) = 160000	(N_8M_8) = 640000
(N_9) = 1	(M_9) = 180000	(N_9M_9) = 180000
$\bullet \sum_1^9 N_i = 80$		$\bullet \sum_1^9 N_iM_i = 8000000$

Number average molecular weight is expressed as

$$\overline{M_n} = \frac{N_1M_1 + N_2M_2 + N_3M_3 + \dots N_iM_i}{N_1 + N_2 + N_3 + \dots + N_i} = \frac{\sum_1^\infty N_iM_i}{\sum_1^\infty N_i}$$

For the above polymer number average molecular weight is $\overline{M_n} = \dfrac{8000000}{80} = 100000$.

Thus the physical significance of the number average molecular weight is the total weight of the polymer divided by the total number of polymer chains. Practically it is not possible to get the information for individual polymer chains those are individual polymer molecules but can get all together.

For calculating weight average molecular weight, you have to understand the weight fraction (W_i) of the polymers. It is the fraction of the total weight of similar chains of the polymers divided by the total weight of the polymers.

Mathematically $W_i = \dfrac{N_iM_i}{\sum_1^\infty N_iM_i}$ and

weight average molecular weight is $\overline{M_w} = \sum_1^\infty W_iM_i$.

Another expression to calculate the weight average molecular weight is

$$\overline{M_w} = \frac{\sum_1^\infty N_iM_i^2}{\sum_1^\infty N_i}.$$

Here $\bullet \sum_1^9 N_iM_i = 8000000$

$\therefore W_1 = \dfrac{20000}{8000000} = 0.0025$

Similarly we can calculate W_2, W_3 ..... W_9.

The previous table can be modified as

(N_i)	(M_i)	(N_iM_i)	$\left(W_i = \dfrac{N_iM_i}{\sum_1^\infty N_iM_i}\right)$	(W_iM_i)
1	20000	20000	0.0025	50
4	40000	160000	0.02	800
8	60000	480000	0.06	3600
16	80000	1280000	0.17	13600
20	100000	2000000	0.25	25000
16	120000	1920000	0.255	30600
8	140000	1120000	0.14	19600
4	160000	640000	0.08	12800
1	180000	180000	0.0225	4050
$\bullet\ \sum_1^9 N_i = 80$		$\bullet\ \sum_1^9 N_iM_i = 8000000$	$\bullet\ \sum_1^9 W_i = 1$	$\bullet\ \sum_1^9 W_iM_i = 110100$

For this polymer weight average molecular weight is 110100. In this regard remember that weight average molecular weight is either equal or higher than number average molecular weight. The ratio of weight average molecular weight and number average molecular weight is called **Polydispersity Index (PDI)**. When PDI equal to 1 then both weight average molecular weight and number average molecular weight are equal and the polymer is called **monodispersed polymer**. In this case PDI is 1.101 and thus it is polydispersed polymer.

$$\text{Polydispersity Index (PDI)} = \frac{\text{weight average molecular weight}}{\text{number average molecular weight}}$$

The expression of Z-average molecular weight is $\overline{M_z} = \dfrac{\sum_1^\infty N_iM_i^3}{\sum_1^\infty N_iM_i^2}$.

The expression of viscosity average molecular weight is

$$\overline{M_v} = \left[\sum_1^\infty W_iM_i^a\right]^{\frac{1}{a}} = \left[\frac{\sum_i^\infty N_iM_i^{a+1}}{\sum_1^\infty N_iM_i}\right]^{\frac{1}{a}}, \text{ where } a \text{ is the hydrodynamic}$$

volume of the polymer. The value of a changes with temperature, solvent and polymer. Usually the value of a varies from 0.5-0.9 but when $a = 1$, $\overline{M_v} = \overline{M_w}$.

The plot of weight fraction against molecular weight of polymers is presented below

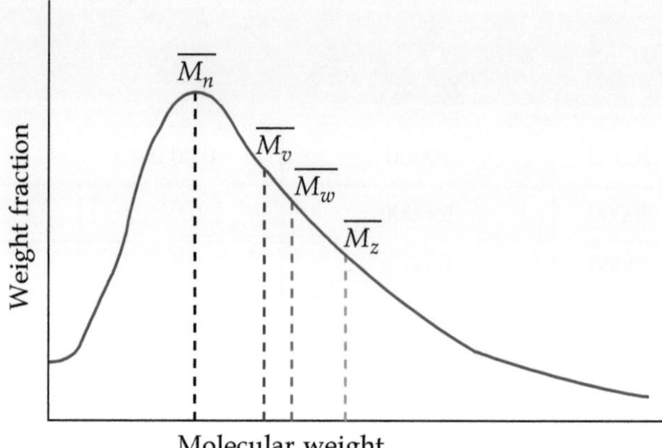

Pic 6.1 Plot of weight fraction against molecular weight of polymers

From the plot it can be predicted that increasing order of molecular weight of polymers are, Z-average molecular weight > weight average molecular weight > viscosity average molecular weight > number average molecular weight.

Significance of molecular weight distribution for the polymer :

If you want to prepare small molecules, each and every time you will get same and definite molecular weight. But for synthetic polymer, if you follow same preparation procedure you will not able to get exact numbers of individual molecular weight of polymer chains. In batch to batch, molecular weight vary little bit. That means you will get different molecular weight distribution for each time of polymer preparation. Different types of distribution are taken according to their own importance. The number average molecular weight ($\overline{M_n}$) calculates the total number molecules in a unit mass of polymer irrespective of their shape or size. The $\overline{M_w}$ calculates the polydispersity of a polymer mixture. $\overline{M_v}$ gives same information like $\overline{M_w}$ i.e. polydispersity index or broadness of molecular weight distribution of polymer. $\overline{M_z}$ is highly sensitive and difficult to measure accurately of that portion of a polymer which have high molecular weight chains. It measure the motion of the polymer molecules.

6.1 Determination of Molecular Weight of Polymers

Polymers' molecular weight can be determined by different methods with different instruments. Remember that all the methods have limitations and instruments have sensitivity for polymers molecular weight determination. Suppose you are measuring a length with an instrument and the instrument can measure the length from 1 cm to 100 cm. Then the lower and higher sensitivity limit of that instrument is 1 and 100 respectively. If you try to measure a distantness below 1 cm or above 100 cm with that instrument, you will get an error. This is true for all instruments whether it measures length, conductivity, viscosity and so on. So before starting any measurement check the sensitivity of the instruments. Every method has also limitations and before selecting any method be careful with its limitations with your experiment. Methods used for molecular weight determination of polymers are relative or absolute or fractionation. Relative methods deal with polymer's physical structure, absolute methods deals with polymers molecular weight and fractionation method deal with polymers chain distribution (effective size). Mass spectrometry, colligative property, end group analysis, light scattering and ultracentrifugation are the examples of absolute method. The example of relative method and fractionation method are solution viscosity and gel permeation chromatography respectively.

Method	Type of average mol. Wt.	Measurement
End group analysis	$\overline{M_n}$	Titration, elemental analysis, spectroscopy
Viscometry	Relative	Ostwald and Ubbelohde viscometer
light scattering	$\overline{M_w}$	Laser light scattering photometer
Osmotic pressure	$\overline{M_n}$	Membrane osmometry, vapour pressure osmometry

In this chapter end group analysis, viscometry, light scattering and osmotic pressure methods for molecular weight determination of polymers are discussed.

6.2 Determination of Molecular Weight of Polymers by End Group Analysis

End groups are the groups that remain at the end of the polymers chain. Generally these end groups have functionality. In a polymer chain the end groups are either of the same type (see AB polymer chain) or different types (see CD polymer chain). From the weight and number of functional groups present (determined by titration) in polymers molecular weight of the polymers can be calculated.

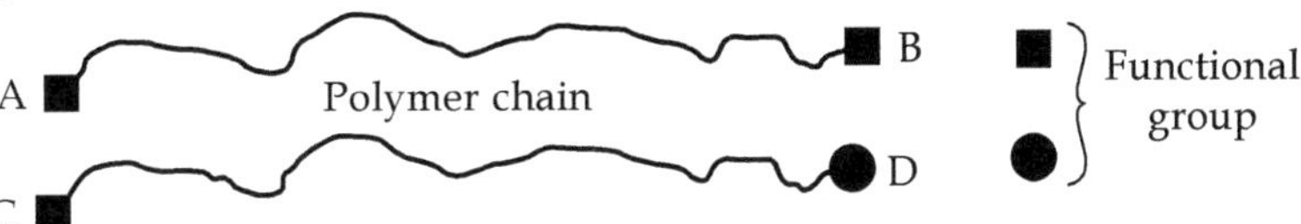

Limitation : This method is suitable for low molecular weight. The end groups should be free for detection. The polymer chains should be linear without branching. There should be not trapped unreacted monomers. The end groups should not interact with each other.

Examples : i) Carboxyl groups in polyethylene carboxylic acid and in polyamides are titrated with base. ii) Amino groups in polyamide are titrated with acid. iii) Hydroxyl groups are reacted with pyromellitic dianhydride (PMDA) and back titrated with NaOH solution.

Here poly(ethylene glycol) (PEG) is taken for discussion. Its viscoelastic property changes with its molecular weight. For molecular weight < 1000 viscous property is higher than elastic property and thus the material is liquid. As its molecular weight increases elastic property increases and becomes solid. Thus for this experiment molecular weight < 1000 is chosen.

Ethylene glycol = HO—CH$_2$CH$_2$—OH

poly(ethylene glycol) = HO—CH$_2$CH$_2$—[O—CH$_2$CH$_2$]$_{n-1}$—OH

Here the end group is —OH (alcohol). Esters are formed by the reaction of this alcohol with PMDA, a carboxylic acid anhydride. The reaction is conducted in N,N-dimethyl formamide (DMF), a non-aqueous solvent and Imidazole, a catalyst is used to speed the reaction. The analysis involves an indirect titration procedure. In previously weighed PEG samples, an excess amount of PMDA and the catalyst is added. After an hour water is added to convert the unreacted anhydride groups to the acid form, and the acid is titrated with standard NaOH to get the number of moles. Thus the number average molecular weight can be calculated from the molarity of PMDA and the number of moles of PEG sample. (for detailed discussion see the practical part).

6.3 Determination of Molecular Weight of Polymers by Viscometry

You are already familiar with viscosity measurement by Ostwald viscometer. There is another apparatus for the measurement of viscosity of solution named Ubbelohde viscometer. The physical construction of this apparatus is little bit different than Ostwald viscometer. As this method is cheaper than other methods for molecular weight determination thus this method is extensively used than others. It is based on the determination of intrinsic viscosity (η) of a polymer in solution through the measurements of solution viscosity. The fundamental relationship between intrinsic viscosity (η) and molecular-weight ($\overline{M_v}$) is

$[\eta] = K\overline{M_v}^a$, where ($\overline{M_v}$) is the viscosity-average molecular-weight, K and a are Mark-Houwink constants for specific polymer and solvent. The equation is also known as **Mark-Houwink-Sakurada equation**. In this regard you have to know few viscosity terminology

Relative viscosity (η_{rel}) = $\dfrac{\eta}{\eta_0}$ = $\dfrac{t}{t_0}$, where η is solution viscosity, η_0 is solvent viscosity, t is flow time of solution and t_0 is flow time of solvent.

Specific viscosity $(\eta_{sp}) = \dfrac{\eta - \eta_0}{\eta_0} = \dfrac{t - t_0}{t_0} = \eta_{rel} - 1$

Reduced viscosity $(\eta_{red}) = \dfrac{\eta_{sp}}{c} = \dfrac{\eta_{rel} - 1}{c}$

Intrinsic viscosity $(\eta) = \left(\dfrac{\eta_{sp}}{c}\right)_{c \to 0}$

Thus when η, K and a are known then $\overline{M_v}$ will be calculated from

$$[\eta] = K\overline{M_v}^{\,a} \quad \text{Or} \quad \log[\eta] = \log K + a \log \overline{M_v}$$

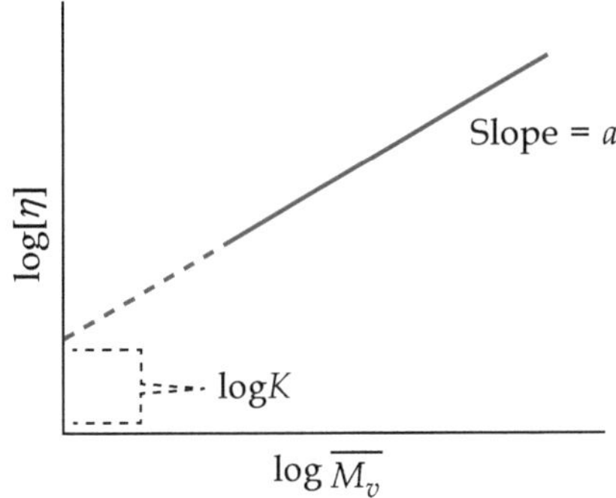

Pic 6.2 Plot of $\log[\eta]$ against $\log \overline{M_v}$ of polymer solutions.

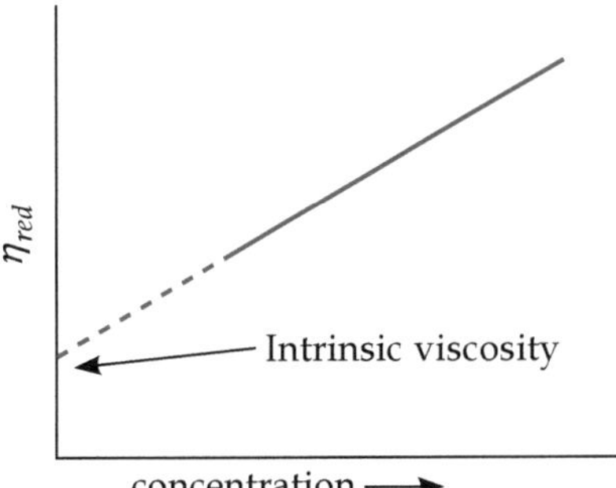

Pic 6.3 Plot of η_{red} against the concentration of polymer solutions.

6.4 Determination of Molecular Weight of Polymers by Static Light Scattering Method

Light has reflection, refraction, diffraction and scattering properties. Scattering of light is a phenomenon of light where it deviates from a straight path on striking molecules. Reflection, refraction or diffraction originate when light is being scattered by ordered particles. Static light scattering method is used to determine the weight average molecular weight. The polymer is dissolved in a solvent and then placed on the path of a laser beam. Depending on the polarity, chain size, and concentration of the polymer sample, light scatters at different angles in respect of the incident beam. By measuring Raleigh scattering of polymer solutions, molecular weight of polymer was determined.

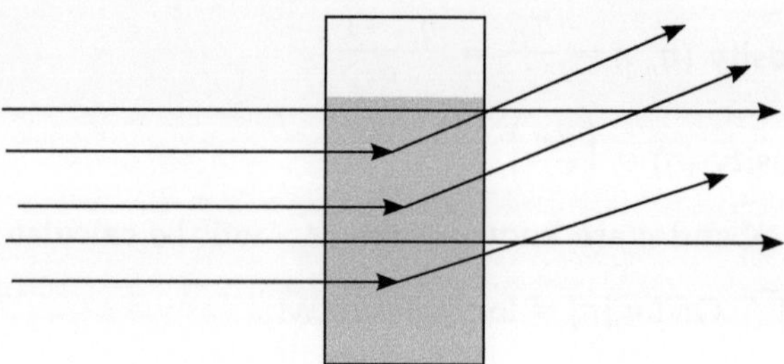

Pic 6.4 Scattering of light from polymer solution

Intensity of scattering light, $I = I_0 e^{-tx}$, where I_0 is the intensity of incident ray and x, t are the thickness and turbidity of the polymer sample. The molecular weight M and the weight concentration C of the polymer solution are related with scattered intensity as

$$I = KCM$$

where K is constant. Concentration should be sufficiently low. Think about dependency of K, it depends on parameters both from the light scattering instrument and polymer solution. Thus the equation can be written as

$$I = K_1 I_0 \frac{V(\theta)}{r^2} CM$$

where $V(\theta)$ is the scattering volume as seen from the detector angle θ and r is the distance between the scattering molecules and the detector. K_1 is constant, that can be calculated from the equation

$$K_1 = \frac{(2\pi n_0)^2 \left(\frac{dn}{dC}\right)^2}{N_A \lambda^4}$$

where n_0 is the refractive index of the pure solvent, $\frac{dn}{dC}$ is the refractive index increment of the polymer solution/pure solvent system, λ is the wavelength of the laser used and N_A is Avogadro's number.

Now a correction factor $P(\theta)$ has been included in the scattered intensity equation for size elimination. Then the equation becomes

$$I = K_1 I_0 \frac{V(\theta)}{r^2} CMP(\theta)$$

$P(\theta)$ has a value 1 at $\theta = 0^0$, otherwise it is always less than one. This function is called the form factor of the molecules which are related to the size and the shape in solution. By simplifying the above equation

$$\frac{I}{I_0} \cdot \frac{r^2}{V(\theta)} = K_1 CMP(\theta)$$

The quantity on the left side in the above equation is called the **Rayleigh ratio R(θ), then**

$$R(\theta) = K_1 CMP(\theta) \quad \text{or} \quad \frac{K_1 C}{R(\theta)} = \frac{1}{MP(\theta)}$$

Considering the influence of polymer solution (non-ideal) concentration on both side of the above equation which is experimentally workable can be obtained by adding an extra term

$$\frac{K_1 C}{R(\theta)} = \frac{1}{MP(\theta)} + 2A_2 C$$

where A_2 is the second virial coefficient, that measures the interaction forces between the dissolved molecules. When A_2 is positive the intermolecular forces are repulsive and when negative, the intermolecular forces are attractive. If the second virial coefficient is zero the solution is called **ideal** and there are no net interactions between the dissolved molecules. Thus to determine the molecular weight of polymer solution it is necessary to measure the scattered intensity at many different angles and with different concentrations.

Limitation: This method is suitable for molecular weight in the region 10^3-10^9 g/mol. The polymer solution must be completely transparent and should not absorb light of the wavelength used. The volume and concentration of polymer solutions should be sufficient (relatively higher concentration and more volume required). The polymer molecules in solution should have different refractive index than that of the solvent.

6.5 Determination of Molecular Weight of Polymers by Osmometry Method

Colligative properties are the property of solution which depends on total number of solute and solvent molecules. Vapour pressure lowering, elevation of boiling point, depression of freezing point and osmotic pressure are the examples of colligative properties. These properties are typically seen in dilute solutions. There are the relations of these colligative properties with molecular weight. Vapour pressure osmometry and membrane osmometry are the two osmometry techniques used for the determination of number average molecular weight. Vapour pressure osmometry is used for lower molecular weight samples where as membrane osmometry can be used for relatively higher molecular weight samples. When a pure solvent is separated from a solution through a semipermeable membrane, the solvent molecules will move through the membrane to the solution. This phenomenon is called **osmosis** and the pressure required to stop the flow of solvent molecules through the semipermeable membrane is called

osmotic pressure. The Van't Hoff equation for the osmotic pressure of an ideal solution is

$$\pi M = cRT$$

where π is the osmotic pressure, M is molar mass, c is concentration in mass per unit volume, R is gas constant and T is absolute temperature.

Polymer solutions are not ideal. McMillan–Mayer modified Van't Hoff equation for the osmotic pressure of polymer solution as

$$\frac{\pi}{c} = \frac{RT}{M_n} + Bc + Dc^2$$

where c is the concentration of the polymer in gram per unit volume of the solution, B and D are virial coefficients expressed in mL mol/g^2. For dilute polymer solution c is small and higher powers of c are always neglected. Then the equation becomes

$$\frac{\pi}{c} = \frac{RT}{M_n} + Bc$$

Thus the molecular weight of the polymer can be calculated from the intercept of the plot of $\frac{\pi}{c}$ vs c.

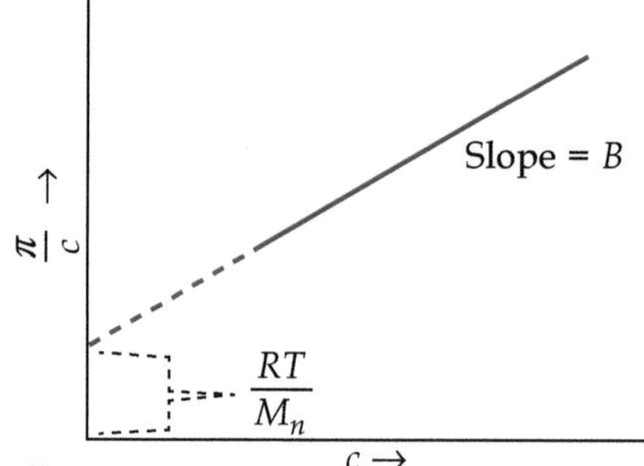

Pic 6.5 Plot of $\frac{\pi}{c}$ against the cocentration of polymer solutions

In vapour pressure osmometry, the lowering of vapour pressure is measured using two sensitive thermistors which are placed in a chamber of saturated solvent vapour. The molecular weight is measured from the difference in resistance (ΔR) of the two thermistors. This method is useful for precisely known low molecular weight.

1. **What is number average molecular weight and weight average molecular weight? How they are related with each other?**

Ans. Polymer is made by polymer chains. Each polymer chain may or may not have the same monomer units. Experimentally, the molecular weight of each polymer chain differs from other polymer chains. Thus the molecular weight of polymer is not a definite number unlike simpler pure compounds and to express its average values of each polymer chains are needed.

Number average molecular weight is expressed as

$$\overline{M_n} = \frac{N_1M_1 + N_2M_2 + N_3M_3 + \dots N_iM_i}{N_1 + N_2 + N_3 + \dots + N_i}$$

Where N_1 polymer chains have molecular weight M_1, N_2 polymer chains have molecular weight M_2 and so on.

Weight fraction (W_i) of the polymer is the fraction of the total weight of similar chains of the polymers divided by the total weight of the polymers.

Mathematically $W_i = \dfrac{N_iM_i}{\sum_1^\infty N_iM_i}$ and weight average molecular weight is

$$\overline{M_w} = \sum_1^\infty W_iM_i = \frac{\sum_1^\infty N_iM_i^2}{\sum_1^\infty N_i}$$

The relation between the number average molecular weight and the weight average molecular weightis that weight average molecular weight is greater than number average molecular weight.

Weight average molecular weight = polydispersity index (PDI) ×

number average molecular weight

When PDI is equal to 1 then both weight average molecular weight and number average molecular weight are equal.

2. Plot weight fraction vs. molecular weight of polymer and point out M_n, M_w, M_z, and M_v.

Ans. The plot of weight fraction against molecular weight of polymer is presented below

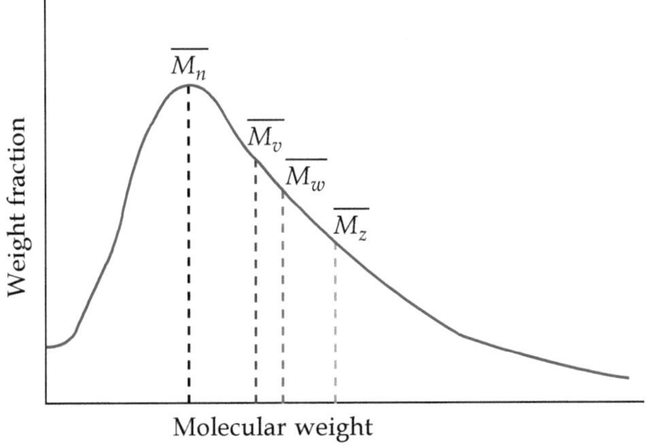

It is clear from the plot that increasing order of molecular weight of polymer is Z-average molecular weight > weight average molecular weight > viscosity average molecular weight > number average molecular weight.

3. Write down the expression out M_n, M_w, M_z, and M_v.

Ans. The expressions are

$$\overline{M_n} = \frac{N_1 M_1 + N_2 M_2 + N_3 M_3 + \dots N_i M_i}{N_1 + N_2 + N_3 + \dots + N_i}$$

$$\overline{M_w} = \sum_1^\infty W_i M_i = \frac{\sum_1^\infty N_i M_i^2}{\sum_1^\infty N_i}$$

$$\overline{M_z} = \frac{\sum_1^\infty N_i M_i^3}{\sum_1^\infty N_i M_i^2}$$

$$\overline{M_v} = \left[\sum_1^\infty W_i M_i^a \right]^{\frac{1}{a}} = \left[\frac{\sum_i^\infty N_i M_i^{a+1}}{\sum_1^\infty N_i M_i} \right]^{\frac{1}{a}}$$

Where N_1 polymer chains have molecular weight M_1, N_2 polymer chains have molecular weight M_2 and so on and a is the hydrodynamic volume of the polymer.

4. Describe Polydispersity Index (PDI) and give an example where its value is 1.

Ans. The ratio of weight average molecular weight and number average molecular weight is called Polydispersity Index (PDI). When PDI equal to 1 then both weight average molecular weight and number average molecular weight have the same value and the polymer is monodispersed.

$$\text{Polydispersity Index (PDI)} = \frac{\text{weight average molecular weight}}{\text{number average molecular weight}}$$

Natural polymers like cellulose have PDI value 1.

5. Write down the Mark-Houwink-Sakurada equation. How will you calculate the molecular weight of a polymer from this equation?

The fundamental relationship between intrinsic viscosity (η) and molecular-weight ($\overline{M_v}$) is $[\eta] = K\overline{M_v}^a$ where ($\overline{M_v}$) is the viscosity-average molecular-weight, K and a are Mark-Houwink constants for specific polymer and solvent. The equation is also known as **Mark-Houwink-Sakurada equation**.

If we know the reduced viscosity of polymer solution having different concentrations, we can calculate intrinsic viscosity from the graph. The intercept will give the value of intrinsic viscosity.

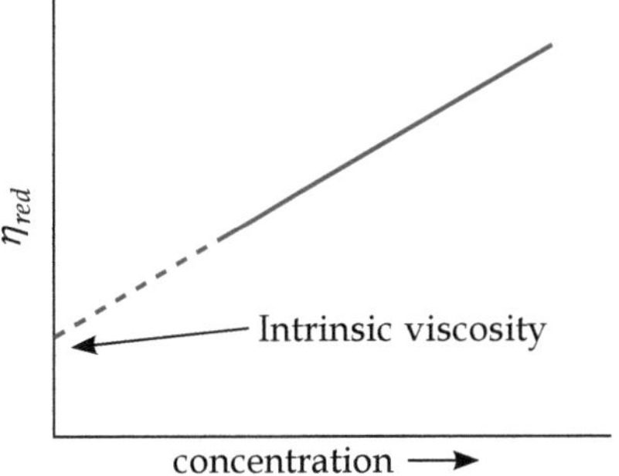

Plot of η_{red} against the concentration of polymer solutions.

when η, K and a is known then $\overline{M_v}$ will be calculated from the following equation

$$[\eta] = K\overline{M_v}^a \quad \text{Or,} \quad \log[\eta] = \log K + a\log \overline{M_v}$$

To make the following plot, polymer solutions having different viscosities have to be measured with respect to their molecular weight. Then for polymer solution having unknown molecular weight can be calculated on measuring the intrinsic viscosity and matching with the plot.

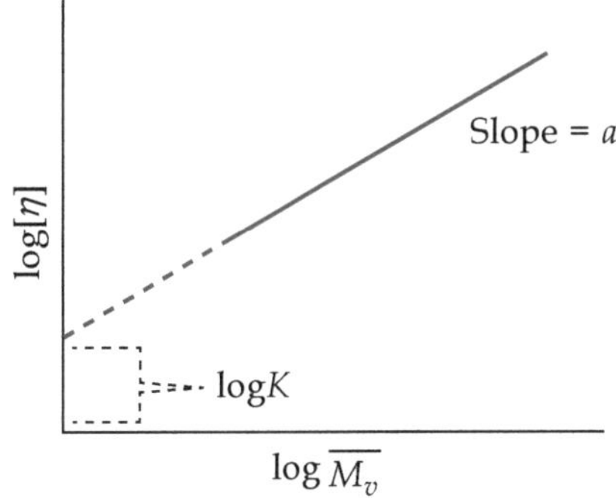

Plot of $\log[\eta]$ against $\log \overline{M_v}$ of polymer solutions.

Exercise

A **Multiple Choice Type Questions(MCQ) (Each Question carries 1 mark)**

1. Polymers are mixtures of molecules of various molecular weights and molecular sizes. Therefore, we must use

   a. averages molecular weight for these substances

   b. mean molecular weight for these substances

   c. mode molecular weight for these substances

   d. none of these

2. Which is true from the following?

   a. $M_z > M_w < M_n$ **b.** $M_z > M_w > M_n$

   c. $M_z < M_w > M_n$ **d.** none of the above

3. Which is true from the following?

   a. $M_z < M_w > M_v > M_n$ **b.** $M_z > M_w < M_v > M_n$

   c. $M_z > M_w > M_v < M_n$ **d.** $M_z > M_w > M_v > M_n$

4. What is the unit of molecular weight?

   a. Kilogram per mole   **b.** Kilogram

   c. mole   **c.** all of these

5. Polydispersity Index (PDI) is the ratio of

   a. M_v and M_n **b.** M_z and M_n

   c. M_w and M_n **d.** M_w and M_z

6. The expression of Z-average molecular weight is

   a. $\overline{M_z} = \dfrac{\sum_1^\infty N_i M_i^3}{\sum_1^\infty N_i M_i^2}$ **b.** $\overline{M_z} = \dfrac{\sum_1^\infty N_i M_i^2}{\sum_1^\infty N_i M_i^2}$

   c. $\overline{M_z} = \dfrac{\sum_1^\infty N_i M_i^3}{\sum_1^\infty N_i M_i^3}$ **d.** $\overline{M_z} = \dfrac{\sum_1^\infty N_i M_i^3}{\sum_1^\infty N_i M_i^3}$

7. The expression of viscosity average molecular weight is

   a. $\overline{M_v} = \left[\sum_1^\infty W_i M_i^a\right]^{\frac{1}{a}}$ where a is the hydrodynamic volume of the polymer

   b. $\overline{M_v} = \left[\sum_1^\infty W_i M_i^{a+1}\right]^{\frac{1}{a}}$ where a is the hydrodynamic volume of the polymer

   c. $\overline{M_v} = \left[\sum_1^\infty W_i M_i^a\right]^{a}$, where a is the hydrodynamic volume of the polymer

   d. $\overline{M_v} = \left[\sum_1^\infty W_i M_i^a\right]^{\frac{2}{a}}$, where a is the hydrodynamic volume of the polymer

8. Titration is used to measure the molecular weight of polymer for

   a. viscometry method   **b.** end group analysis method

   c. light scattering method   **d.** osmometry method

9. The fundamental relationship between intrinsic viscosity (η) and molecular-weight ($\overline{M_v}$) is (where K and a are Mark-Houwink constants for specific polymer and solvent)

   a. $[\eta] = K\,\overline{M_v^{a+1}}$ **b.** $[\eta] = aK$

   c. $[\eta] = K\,\overline{M_v}^{a}$ **d.** None of these

10. The molecular weight M and the weight concentration C of the polymer solution related with scattered intensity as (where K is constant)

   a. $I = KCM$ **b.** $I = KCM^2$

   c. $I = KC^2M$ **d.** $I = KC^2M^2$

B **Short Answer Type Questions (Each question carry either 2 or 3 marks)**

Write short notes on the following

1. Explain each terms M_n and M_w. Mention them in a single Plot.
2. Explain each terms M_n and M_z. Mention them in a single Plot.
3. Explain each terms M_n and M_v. Mention them in a single Plot.
4. Explain each terms M_w and M_v. Mention them in a single Plot.
5. Describe the Viscometry method for the determination of polymers molecular weight.
6. Describe the light scattering method for the determination of polymers molecular weight.
7. Describe the osmometry method for the determination of polymers molecular weight.
8. How will you calculate the monomer molecular weight (unreacted) in a polymer?
9. What are the reasons of increasing molecular weight of some chemicals (polymers or macromolecules) in plant when the concentration of a pollutants are more that effect the plant?

C **Long Answer Type Questions (Each question carry 5 marks)**

1. Explain each terms M_n, M_w, M_z, and M_v. Plot them.
2. Write down two methods elaborately for the determination of molecular weight of polymers.
3. Describe the end group analysis method for the determination of polymers molecular weight.
4. A polyamide was prepared by bulk polymerization of hexamethylene diamine (11.525 g) with adipic acid (14.6 g) at 300°C. Analysis of the whole reaction product showed that it contained 2.0×10^{-3} mol of carboxylic acid groups. Evaluate the number-average molar mass, M_n, of the polyamide, and also estimate its weight-average molar mass, M_w, by assuming that it has the most probable distribution of molar mass.
5. A 1 L of polymer solution contains 10 g of a polymer. A non-solvent was added to the solution in steps and different fractions of polymer samples were precipitated out in each step of non-solvent addition. The polymer fractions were washed, dried, weighed, and subjected to M_n determination. Assuming that for each fraction $M_n = M_w$ calculate the M_n, M_w, and PDI of the original polymer sample that was dissolved at the beginning.

Fraction No.	Weight (g)	M_n (= M_w) (g mol^{-1})
1	1.0	2,000
2	5.0	20,000

Fraction No.	Weight (g)	M_n (= M_w) (g mol^{-1})
3	20.0	50,000
4	5.0	100,000
5	1.0	500,000

6. For a polydisperse polymer sample rank the M_n, M_w, M_z, and M_v according to their values. In 10 g of a polystyrene sample (M_n 30,000; M_w 60,000), you add 1 g of a mono-disperse polystyrene sample of (i) $M_w = M_n = 10,000$ (ii) $M_w = M_n = 30,000$ (iii) $M_w = M_n = 45,000$ (iv) $M_w = M_n = 60,000$ (v) $M_w = M_n = 80,000$. What would be the M_w and M_n for new mixed samples ? Lower or higher than the first sample? Between M_w and M_n which one will change more (in terms of %)?

Answers

A

1. a   **2.** b   **3.** d   **4.** a   **5.** c   **6.** a   **7.** a   **8.** b   **9.** c   **10.** a

●———●

7

Glass Transition Temperature of Polymers

 Free Volume and T_g

Every matter in the universe occupies space, called volume. Atoms are not the smallest unit of matter. Till now quarks is the smallest particle but there may be something even smaller. Whatever be the smallest thing, when it moves individually there is free space or volume. Suppose P is the smallest unit of particle, whether it is in an open system or closed system if it moves then there is free volume. See the following image for the understanding of free volume. Free volume also arises when more P stay together. It is important that to move a matter free volume is required. Free volume is the unoccupied space of the matters.

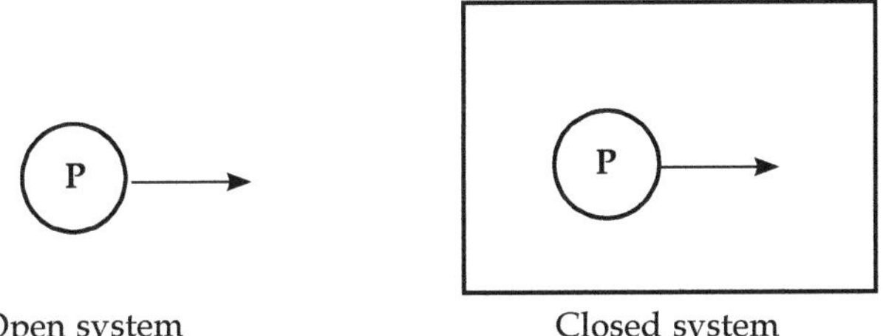

Open system   Closed system

Now consider polymer which consists of many polymer chains (molecules). Powder, sheet, granule whatever be the shape of polymer, it is not formed by a single polymer chain, it contains a lot of polymer chains. The polymer chains may be straight or entangled. See the following figures to understand the polymer chain entanglement.

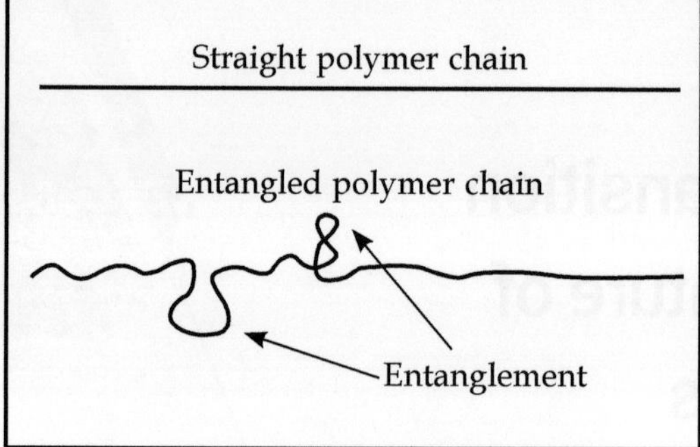

Pic 7.1 Polymer chain entanglement

There is space when a polymer chain is entangled or end which originates free volume. The following figure presents how is free volume formed in the polymer. By combining all these free volumes, the total polymer free volume will be obtained. A polymer chain will move or not is directed by the free volume. If there are more free volume, the polymer chains can easily move or vice versa. Thus free volume controls the polymer chains mobility which in turn influence the physical and mechanical properties of the polymer.

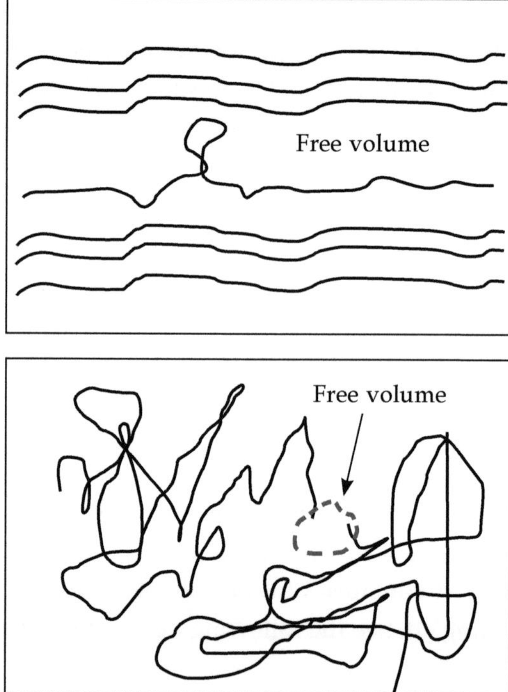

Pic 7.2 Free volume of polymer

Specific volume of a material is the volume occupied in cubic meters (m^3) by its mass measured in kilograms. Most of the polymers have very interesting behaviour, depending on temperature it can remain in a glassy state, rubbery state or molten state. The volume occupied by the polymer in these three different states are different. The thermal expansion of the polymer in these three different states are also different. Thus when specific volume is plotted against temperature (dilatometric study), a transition is obtained. It is a second order transition called glass transition temperature or T_g. (T_g : The glass transition temperature is the temperature below which the polymer behaves as glass (rigid) and above which the polymer behaves as rubber (soft).) See the following figure.

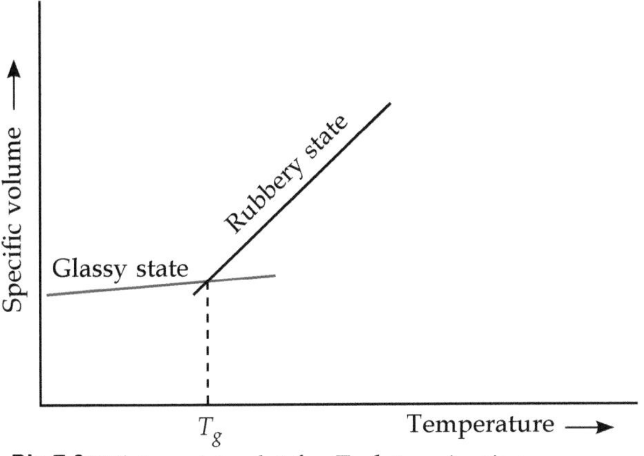

Pic 7.3 Dilatometric plot for T_g determination

Eyring was the first scientist who dealt with the concept of free volume. This dilatometric plots are also used to calculate the free volume.

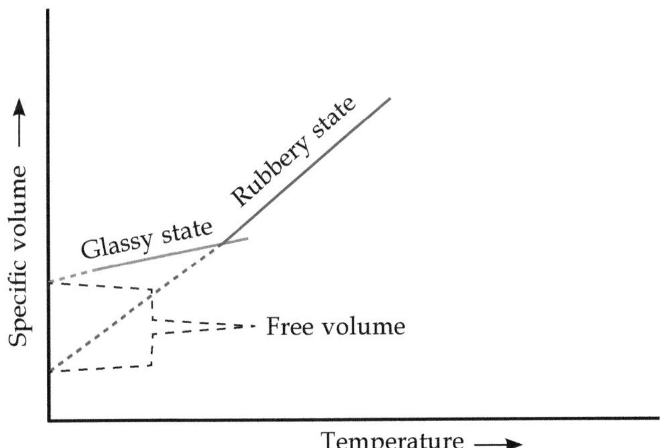

Pic 7.4 Dilatometric plot of free volume calculation by Simha and Boyer

It is the difference in volume of glassy state and a melted state for the polymer at absolute zero temperature. Experimentally this measurement is not feasible and thus extrapolation is needed. It was first postulated by Simha and Boyer that at a fixed value of free volume the glass transition occurs.

7.2 Factors Affecting Glass Transition Temperature (T_g)

The factors which influence T_g of polymers have depends on the below parameters. All the polymers have different structures including conformational, geometrical, chain length, side chain and so on. Each and every structure signifies change in free volume which changes the T_g. Thus the parameters are

1. Structural conformation.

2. Geometrical conformation.

3. Polymer chain length (molecular weight).

4. Side chain (length and substitution).

5. Fillers (example: clay, CNT, metal nanoparticles etc).

7. Plasticizer.

8. Processing and measurement.

Glass transition temperatures of polymers :

Polymer	Glass transition temperature (T_g)
Polyethylene	–80°C
Polypropylene	–10°C
Poly (1,2-butadiene)	–13°C
Poly(1,4-butadiene)	–105°C
PEG	–67°C
PTHF	–86°C
PMMA	105°C
PEMA	65°C
Poly(hexadecyl methacrylate)	16°C
PAN	109°C
Polystyrene	100°C
PVA	80°C
PVAc	33°C
PVC	83°C
PVF	52°C
PVDF	–34°C

Polymer	Glass transition temperature (T_g)
PTFE	119°C
Neoprene	–36°C
Nylon 6,6	51°C
Nomex	264°C
Kevlar	327°C

7.3 Determination of T_g

T_g of polymers can be determined from a number of instruments with different concepts. Here only plots of different experiments are presented to understand the determination of Tg. For detailed discussion see practical book on polymers. The plots are as follows

1. Specific volume measurement.
2. Heat capacity measurement.
3. Storage modulus measurement.
4. Loss modulus measurement.
5. Tan delta measurement.
6. Permittivity measurement.

Specific volume measurement

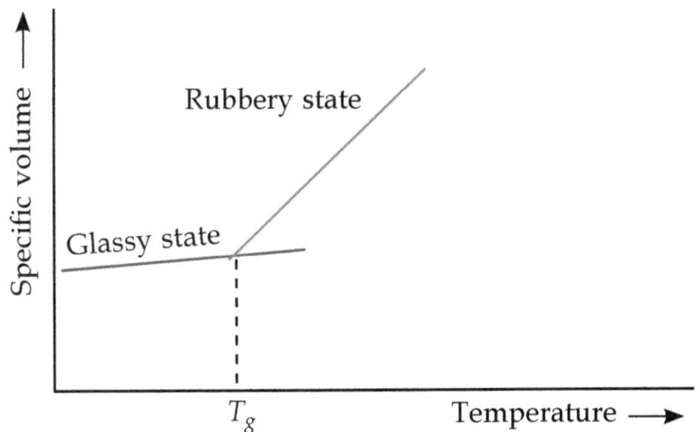

Pic 7.5 Specific volume plot for T_g determination

Heat capacity measurement

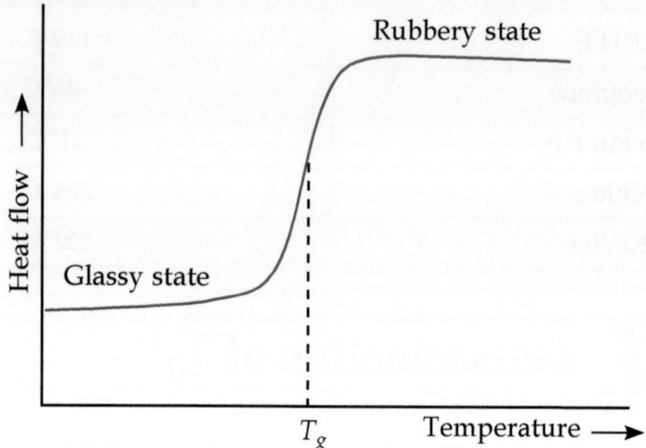

Pic 7.6 DSC plot for T_g determination

Storage modulus measurement

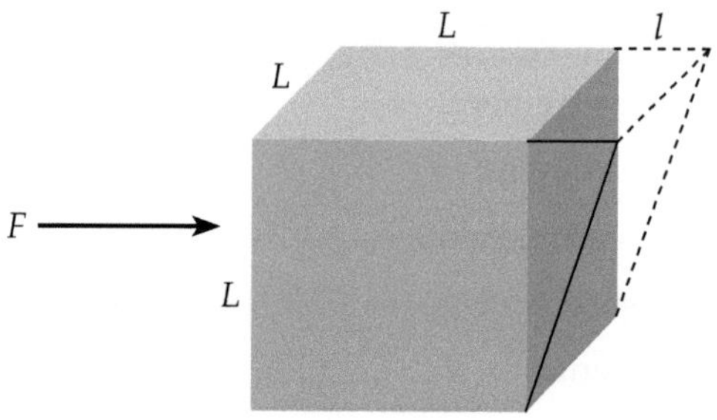

$$\text{Stress} = \frac{\text{force}}{\text{area}} = \frac{F}{L^2}$$

$$\text{Strain} = \frac{\text{length extension}}{\text{original length}} = \frac{l}{L}$$

$$\text{Modulus} = \frac{\text{stress}}{\text{strain}}$$

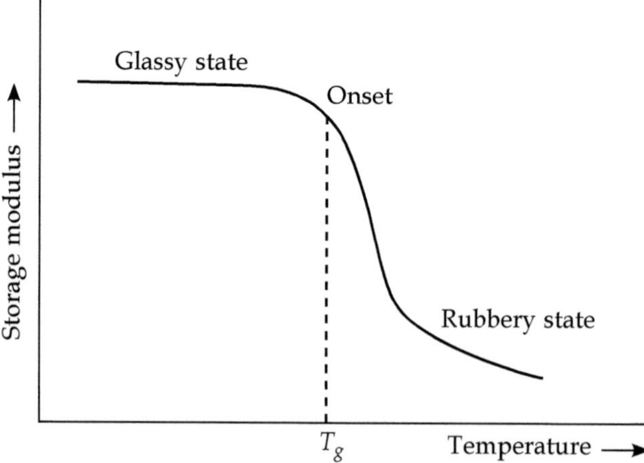

Pic 7.7 DMA plot of storage modulus against temperature for Tg determination

Loss modulus measurement

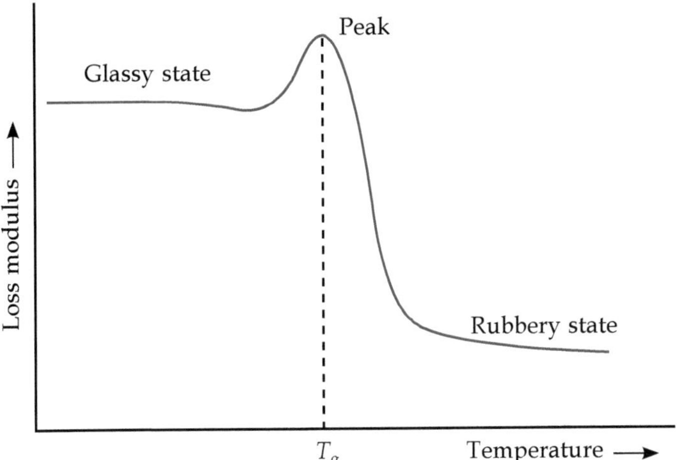

Pic 7.8 DMA plot of loss modulus against temperature for Tg determination

Tan delta measurement

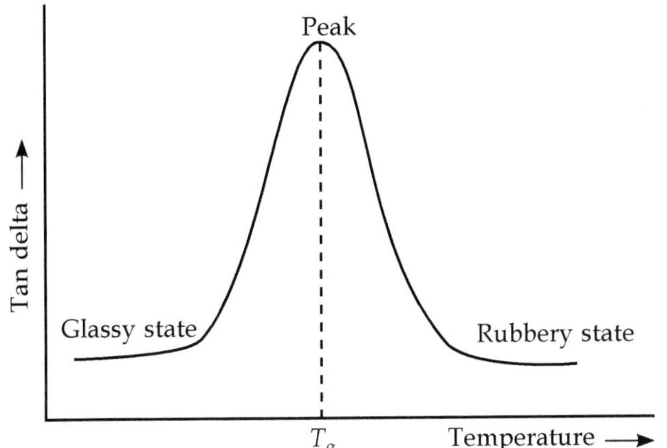

Pic 7.9 DMA plot of tan delta against temperature for T_g determination

Permittivity measurement

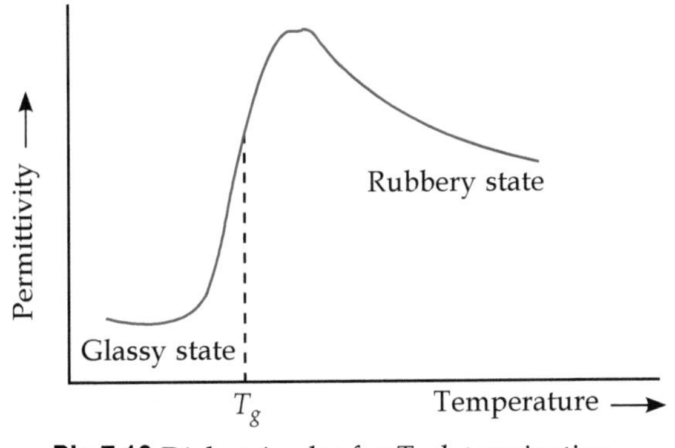

Pic 7.10 Dielectric plot for T_g determination

7.4 WLF Equation

This equation is basically formed by considering the free volume and viscous property of the polymer. In the real world neither any material is perfectly elastic nor perfectly viscous. For perfectly elastic material, if external force is removed then the material will come back to its original shape and size. But in reality, no materials show such behaviour and quartz fibre and Phosphor bronze are the examples of nearly perfect elastic materials. Viscoelastic materials show both elastic and viscous behaviour simultaneously. The term viscous is related to fluidic property of a material. As far we know the viscosity is strongly related with temperature. But it also varies with time of contact. Example is silly putty, silicone polymers which flow like liquid and also breakable into pieces. For short time contact it behaves as an elastic solid and for long period contact it behaves as a viscous liquid. Thus viscosity depends on both temperature and time indicates there is a correlation between time and temperature. **Malcolm L. Williams, Robert F. Landel** and **John D. Ferry** formed an equation to determine temperature-dependent mechanical properties of linear viscoelastic materials which is known as **WLF equation**. The polymer is tested at few different temperatures where the nature of curve remains same but with changing temperatures it only shifts left or right. By combining all the experiments, a master curve at a given temperature is formed. This master curve is used to predict the mechanical properties of the material at a temperature other than those temperatures for which the material tested.

Dynamic viscosity is expressed as shear stress divided by shear rate. In this respect fluid can be classified into **Newtonian fluid** and **non-Newtonian fluid**. See the following figure to understand the nature of fluid.

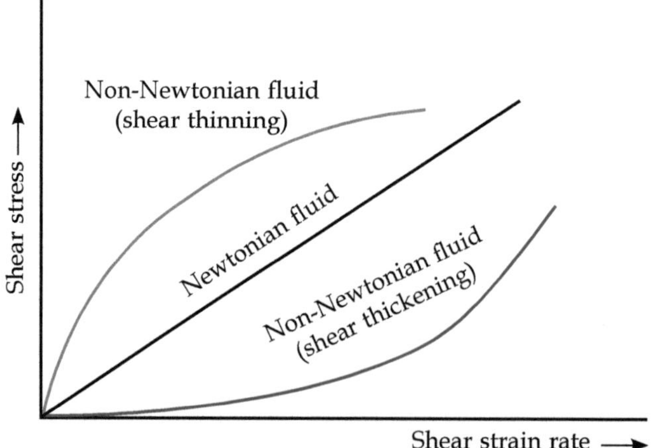

Pic 7.11 Shear stress vs. shear strain rate plot for Newtonian fluid and non-Newtonian fluid

Only Newtonian liquids having low molecular weight polymers follow Arrhenius equation for viscosity

$$\eta = Ae^{\frac{E}{RT}},$$

where η, A, E, R and T are viscosity, material constant, activation energy, gas constant and the absolute temperature respectively.

By taking log the equation becomes

$$\log \eta = \log A + \frac{E}{RT}$$

Doolittle equation for entangled polymer systems

$$\log \eta = \log A + B\left(\frac{V_0}{V_f}\right)$$

where B is constant, V_0 and V_f are the occupied volume and free volume respectively.

Let P is the probability of reptile motion of polymer chain per unit time then with the help of Arrhenius relationship involving an Eyring activation energy it can be written as

$$P = Ae^{-\left(\frac{E}{kT}\right)}$$

The total probability in time "t" is $P(t)$, then

$$P(t) = tAe^{-\left(\frac{E}{kT}\right)} \quad \text{or,} \quad \ln P(t) = \ln t - \frac{E}{kT}$$

For particular free volume, total motion of polymer chain in time t will be constant *i.e.,*

$$\ln P(t) = \text{constant}$$

Then right-hand side of the equation will also be constant, then

$$\ln t - \frac{E}{kT} = \text{constant} \quad \text{or,} \quad \ln t = \text{constant} + \frac{E}{kT}$$

This is called **time-temperature superposition** where time is inversely related with temperature.

Let $\dfrac{E}{kT} = \dfrac{B}{f}$, where B is constant and f is fractional free volume. Then the above equation becomes

$$\ln t = C + \frac{B}{f},$$

where C is another constant. Now consider two different states at time t_0 and t related with temperature T_0 and T will form two different equation

$$\ln t = C + \frac{B}{f} \quad \text{and} \quad \ln t_0 = C + \frac{B}{f_0}, \text{ substituting these two equation}$$

$$\ln t - \ln t_0 = C + \frac{B}{f} - C - \frac{B}{f_0}$$

$$\ln a_T = \frac{B}{f} - \frac{B}{f_0} = B\left(\frac{1}{f} - \frac{1}{f_0}\right)$$

Again f is related to f_0 as $f = f_0 + \alpha_f(T - T_0)$ where f_0 is the fractional free volume at temperature T_0 and α_f is the expansion coefficient of the free volume. Thus the equation becomes

$$\ln a_T = B\left[\frac{1}{f_0 + \alpha_f(T - T_0)} - \frac{1}{f_0}\right]$$

$$\ln a_T = B\left[\frac{f_0 - f_0 - \alpha_f(T - T_0)}{f_0\{f_0 + \alpha_f(T - T_0)\}}\right]$$

$$\ln a_T = -B\left[\frac{\alpha_f(T - T_0)}{f_0\{f_0 + \alpha_f(T - T_0)\}}\right]$$

Dividing the upper and lower part on right hand side by $f_0\alpha_f$, the equation becomes

$$\ln a_T = -B\left[\frac{\alpha_f(T - T_0)/f_0\alpha_f}{f_0\{f_0 + \alpha_f(T - T_0)\}/f_0\alpha_f}\right]$$

$$\ln a_T = -B\left[\frac{(T - T_0)/f_0}{\{f_0/\alpha_f + (T - T_0)\}}\right]$$

$$\ln a_T = -\frac{\left(\dfrac{B}{f_0}\right)(T - T_0)}{(f_0/\alpha_f) + (T - T_0)}$$

Transforming into log

$$\log a_T = -\frac{B}{2.303 f_0}\left[\frac{(T - T_0)}{(f_0/\alpha_f) + (T - T_0)}\right]$$

$$\log a_T = \left[\frac{C_1(T - T_0)}{C_2 + (T - T_0)}\right]$$

where C_1 and C_2 are $\dfrac{B}{2.303 f_0}$ and $\dfrac{f_0}{\alpha_f}$ can be measured experimentally. It is WLF equation. When T_0 is T_g, the value of C_1 and C_2 are 17.44 and 51.6 respectively which implies that the free volume of a glassy polymer at T_g is 2.5% (2.35% based on recent thermodynamic data) of the total volume.

1. What is free volume and how it originates in polymers?

Ans. Polymer chains have different shapes like linear, entangled, branched or network.

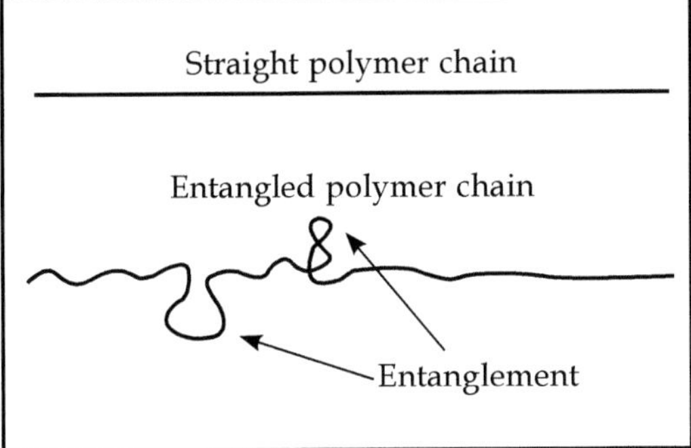

Polymer chain entanglement

There is void space when a polymer chain is entangled and/or when a polymer chainendsand another polymer chain starts. These void spaces originate free volume. The following figure presents how free volume originates in the polymer.

By combining all these free volumes, the total polymer free volume can be calculated. A polymer chain will move or not is directed by the free volume. If there are more free volume, the polymer chains can easily move or vice versa. Thus free volume controls the reptile motion of polymer chains.

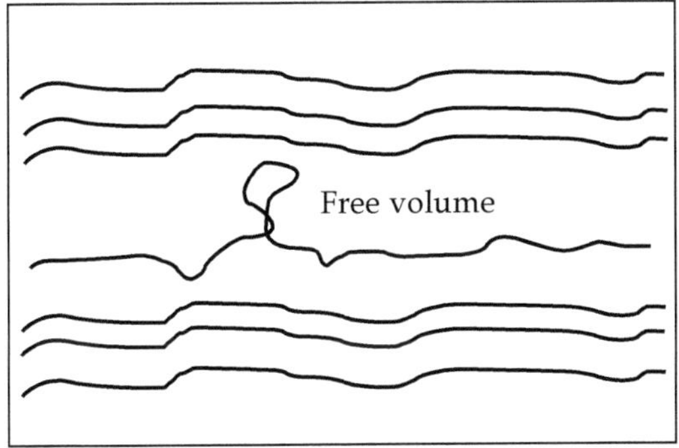

Free volume of polymer originates from chain entanglement

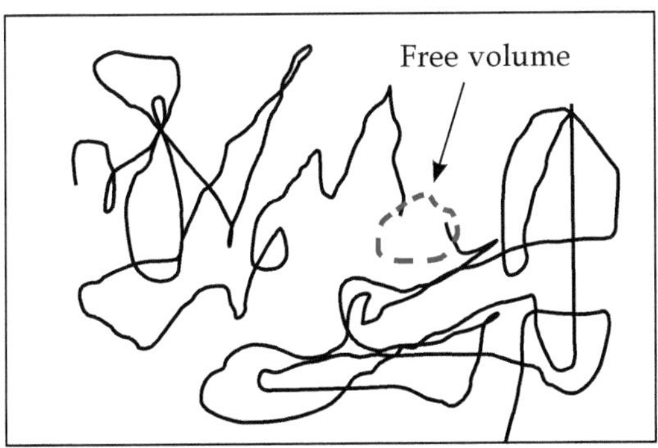

Free volume of polymer originates from
polymer chains end to end spacing

2. Draw the dilatometric plot for T_g determination.

Ans. This is the dilatometric plot for T_g determination of polymer.

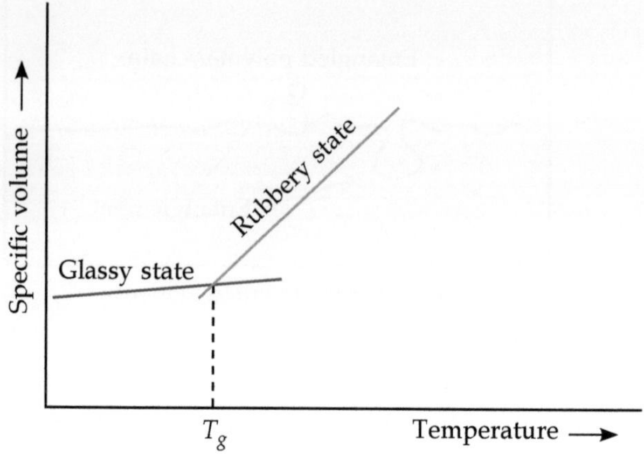

Specific volume of glassy state does not increase or increase very little with temperature. But for the rubbery state this increase is significant. In the plot they will cross each other and the extrapolation line from that point to the x-axis measures the glass transition temperature of that polymer.

3. Draw the DSC plot for T_g determination.

This is the DSC thermogram for T_g determination of polymer.

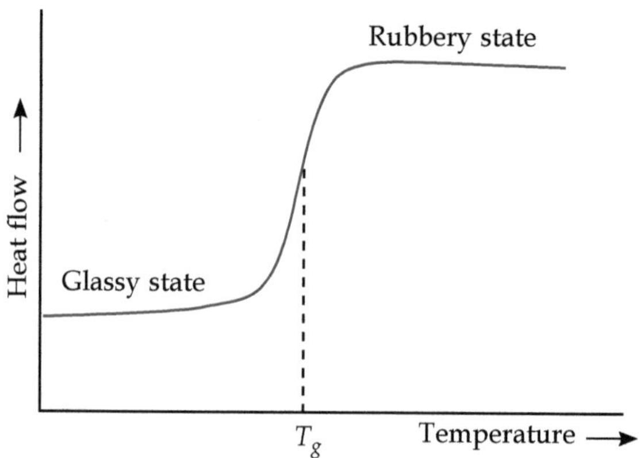

DSC plot for T_g determination

In the DSC thermogram (it is a part of the DSC thermogram for simplicity to understand) it is seen that with increase in temperature heat flow is gradually increasing and then shows a sharp increase and finally constant. From the point where it starts to increase in respect to baseline is called **onset point** and the first point where it becomes constant is called **plateau.** the glass transition temperature is calculated from these two points (onset point and plateau).

4. **What are the methods available for the determination of T_g?**

Ans. T_g of polymers can be determined from specific volume measurement, heat capacity measurement, storage modulus measurement, loss modulus measurement, tan delta measurement and permittivity measurement. There are also other methods for the determination of T_g for polymers. But they are not widely used.

5. **Mention the factors that affect glass transition temperature (T_g) of polymers.**

Ans. The factors which influence T_g of polymers depend on the following parameters

1. Structural conformation.
2. Geometrical conformation.
3. Polymer chain length (molecular weight).
4. Side chain (length and substitution).
5. Fillers (example: clay, CNT, metal nanoparticles etc).
7. Plasticizer.
8. Processing and measurement.

Exercise

A **Multiple Choice Type Questions(MCQ) (Each Question carries 1 mark)**

1. Macromolecule that is not a polymer
- **a.** lipids
- **b.** Proteins
- **c.** Carbohydrates
- **d.** nucleic acids

2. What factors affect glass transition temperature?
- **a.** Pressure and free volume
- **b.** Only pressure
- **c.** Only free volume
- **d.** None of these

3. Which instrument is commonly used to measure T_g?
- **a.** NMR
- **b.** XRD
- **c.** XPS
- **d.** Differential Scanning Calorimetry (DSC)

4. How glass transition temperature of a polymer be increase?
- **a.** ageing time
- **b.** controlling heating rate
- **c.** controlling cooling rate
- **d.** in humidity

5. In glass transition temperature
- **a.** polymer substrate changes from glassy state to rubbery state
- **b.** polymer substrate changes from glassy state to liquid state
- **c.** polymer substrate changes from glassy state to gaseous state
- **d.** none of these

6. Above T_g, the rubbery, flexible polymer has

   a. higher heat capacity   **b.** lower heat capacity

   c. no heat capacity   **d.** none of these

7. Glass transition is

   a. exothermic   **b.** endothermic

   c. First order   **d.** none of these

8. Does crystallinity affect T_g?

   a. no   **b.** yes

   c. may be   **d.** none of these

9. Does degree of polymerization affect the melting point?

   a. yes   **b.** no

   c. may be   **d.** none of these

10. Can we have a 100% crystalline polymer theoretically?

   a. yes   **b.** no

   c. may be   **d.** none of these

B Short Answer Type Questions (Each question carry either 2 or 3 marks)

1. What is free volume and how it affects T_g?

2. Describe dilatometric plot for T_g determination.

3. What are the factors that affect the glass transition temperature?

4. How T_g can be determined from the specific volume measurement?

5. How T_g can be determined from the heat capacity measurement?

6. How T_g can be determined from the heat capacity storage modulus measurement?

C Long Answer Type Questions (Each question carry 5 marks)

1. What are the factors that affect the glass transition temperature? Explain with examples.

2. What are the methods available for the determination of T_g?

3. Derive WLF equation.

4. How free volume is related for the determination of T_g?

Answers

A

1. a	**2.** a	**3.** d	**4.** c	**5.** a	**6.** a	**7.** b	**8.** b	**9.** a	**10.** a

8

Polymer Solution

The constituents of solution are solute and solvent. Depending on the size of the solute molecules in the solvent, solutions are classified into three categories. These are **true solution, colloidal solution** and **suspension**. In true solution, the particle size of solute molecules is less than 1 nm whereas in colloidal solution particle size is in between 1 to 100 nm. For suspension the particle size is more than 100 nm. The solubility of solute molecules in solvent molecules mainly depends on solute-solvent interactions.

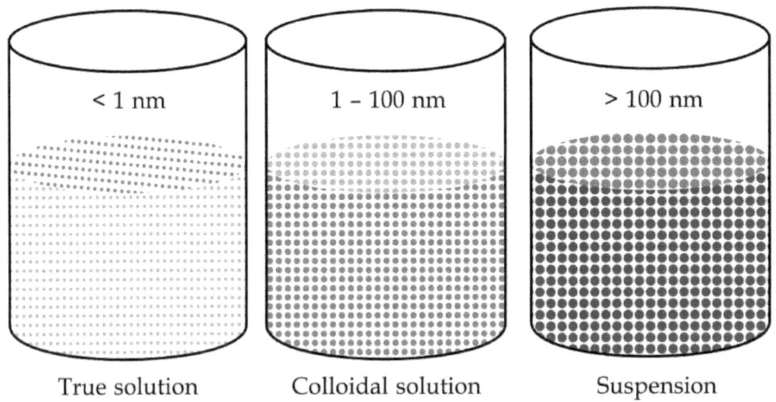

Consider the polymer molecules (size, shape and dimension of polymer molecules) as solute molecules here. More number of interactions steer the polymer molecules (chains) come into solution. If there are no interactions between polymer chains and solvent molecules then the polymer will be insoluble in that solvent. Then the polymer will either remain immersed or float on the solvent depending on the density of both. When the polymer is dissolved in a solvent then the solvent is called good solvent with respect to that polymer. Again when the polymer is precipitated in a solvent then the solvent is called poor solvent with respect to that polymer. Thus a solvent molecule can be classified as good or poor solvent with respect to a polymer. By the way the solute molecules present in the solution may be one or more than one in chemical composition. Similarly the solvent molecules present in the solution may be one or more than one in chemical composition. When two solvent molecules are miscible with each other and present in the same solution, they form a homogeneous solution. But when two solvent molecules are immiscible with each other and present in the same solution, they form a heterogeneous solution. Here the solvents molecules form two different phases, called biphasic solution. In this regard keep in mind that the solution may be multiphasic in nature.

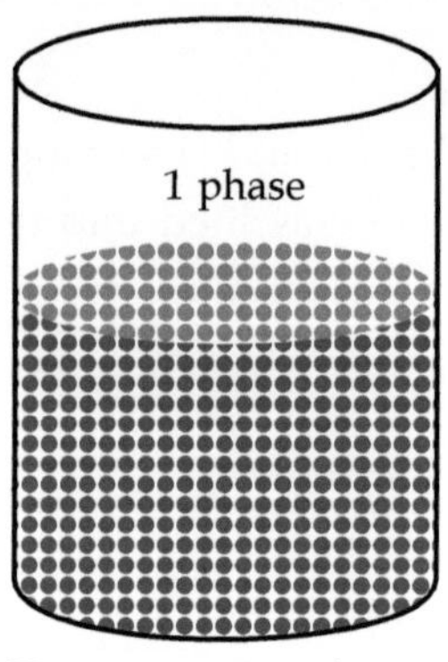

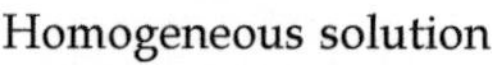

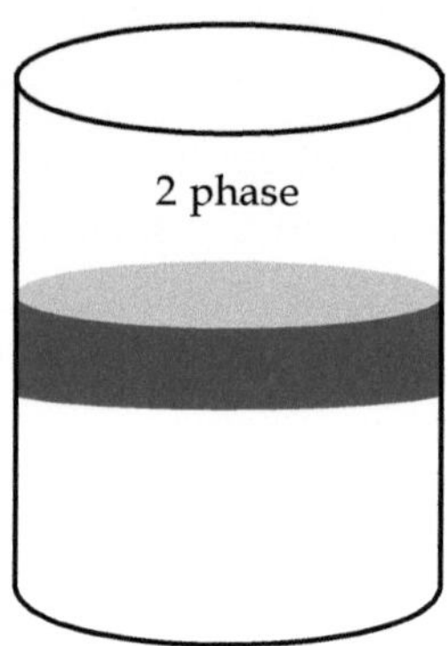

Homogeneous solution　　　　Heterogeneous solution

8.1 Criteria for Polymer Solubility

Solubility of polymers mainly depends on nature of polymer (size, shape and dimension of polymer molecules), molecular weight etc. of polymer as well as the nature of solvent molecule and temperature of the polymeric solution. Smaller the dimension of polymers, greater will be the polymer - solvent interaction, higher will be the chances of polymers becoming soluble. Thus lower molecular weight polymers have more tendency to come into solution (more soluble) than high molecular weight polymers. The interaction mainly will depend on van der Waals forces, hydrogen bonding, polarity, heat of vaporization and cohesive energy density etc. acting on the total system. It is a guessing phenomenon that polar solvent dissolves polar polymers and non-polar solvent dissolves non-polar polymers. In thermodynamic point of view, energy is required when a polymer comes into solution. The energy used will be in the following form namely mechanical shaking, stirring, sonication, vortexing and others form. The solvent molecules penetrate into polymer network structure or inserted into polymer lattice structure or swell the bulk polymer. For this step energy is required and the amount of energy will be higher for ordered structure polymers. Thus dissolving a rubbery polymer will be easier than dissolving a crystalline polymer in terms of energy requirement. When temperature is increased, energy is also increased and thus polymers have higher solubility. But few polymers also show reverse effects.

Polymers solutions are important in terms of polymer synthesis (help to choose proper solvent for the polymer synthesis), polymer processing (film preparation by solvent cast, spin cast or normal evaporation), polymer characterizations (by NMR, UV-Vis, photoluminescence etc.), and its applications (paints, printing ink, spinning etc.).

8.2 Solubility Parameter

The founder of solubility theory, Joel Henry Hildebrand studied the solubility of non-electrolytes and proposed a numerical value (square root of the cohesive energy density) which indicates the solvency behaviour of a specific solvent called **Hildebrand solubility parameter or solubility parameter**. The solubility parameter δ_i of substance i is mathematically expressed as

$$\delta_i = \left(c_i\right)^{\frac{1}{2}} = \left[\frac{E_i}{V_i}\right]^{\frac{1}{2}} = \left[\frac{\Delta H_i - RT}{V_i}\right]^{\frac{1}{2}}$$

where E_i is the molar energy of vaporization. c_i, ΔH_i and V_i are the cohesive energy density, heat of vaporization and molar volume of substance i, R and T are gas constant and temperature respectively. Hildebrand solubility parameter is expressed in mega-Pascal.

Later on, Charles M. Hansen used this theory to predict whether one material will dissolve in another and form a solution.

8.3 Thermodynamics of Polymer Solutions

When a polymer is mixed with solvent, a mixture is formed. Each system (state) has definite enthalpy (H) and entropy (S) at a particular temperature and pressure. According to the Gibbs-Helmholtz equation G (Gibbs energy or Gibbs free energy) of a system is expressed as

$$G = H - TS$$

where T represents the absolute temperature. In a reaction, there are two different states *i.e.*, before reaction and after reaction. The change of G, H and S of these two states are ΔG (change in Gibbs free energy), ΔH (change in enthalpy) and ΔS (change in entropy) respectively. Thus Gibbs free-energy of polymer mixing are expressed as

$$\Delta G_{\text{mix}} = \Delta H_{\text{mix}} - T\Delta S_{\text{mix}}$$

A negative ΔG means that the reactants have more free energy than the products. As the term $T\Delta S_{mix}$ is always negative, the sign of ΔG_{mix} will mainly be determined by ΔH_{mix} (its sign and magnitude). A negative value of ΔG_{mix} indicates that a solvent-polymer system forms a homogeneous solution (polymer and solvent are miscible).

8.4 Models of Polymer Chains

To understand the polymers structure, behaviour and other properties several models of polymers are imagined. There are many models but here few main models are presented.

i. Atomistic model of polyethylene

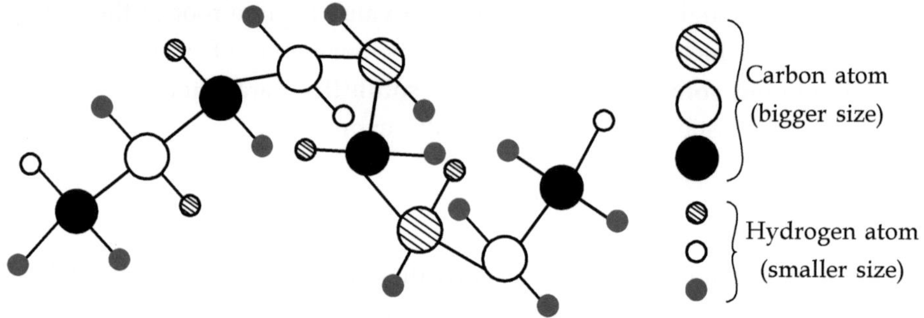

(Solid circle = atom above the plane, empty circle = atom below the plane, pattern circle = atom in the plane.

ii. Main-chain atoms model of polyethylene

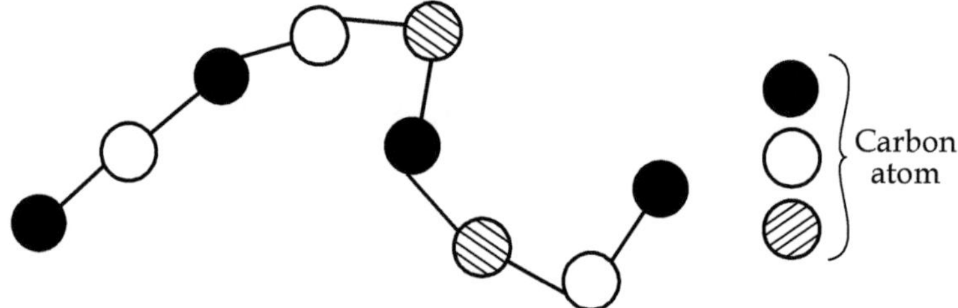

(Solid circle = atom above the plane, empty circle = atom below the plane, pattern circle = atom in the plane)

iii. Bonds model of polyethylene

(Solid bold line = bond above the plane, broken line = bond below the plane, solid line = bond in the plane, bonds are only for main chain)

iv. Thread model of polyethylene

Flexible thread model is the most simplified chain conformation of polymer chain.

v. Bead-stick model of polyethylene

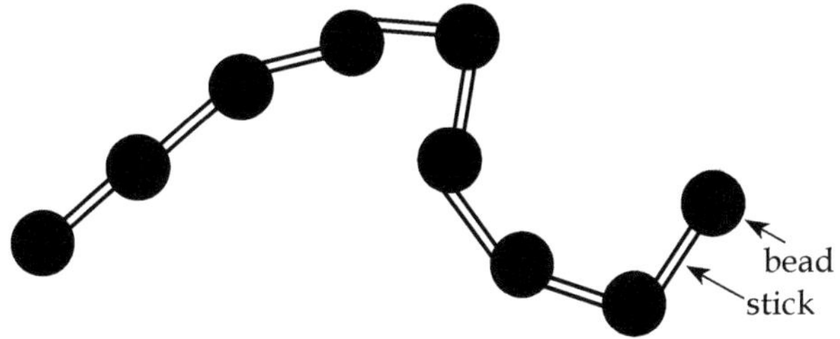

Beads represent polymers backbone molecules and sticks represent bond between backbone molecules.

vi. Bead-spring model of polyethylene

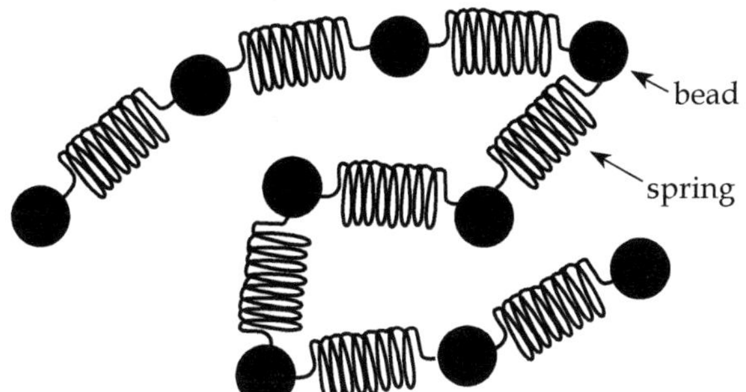

Beads represent polymers backbone molecules and springs represent bond between backbone molecules.

vii. Pearl-necklace model of polyethylene

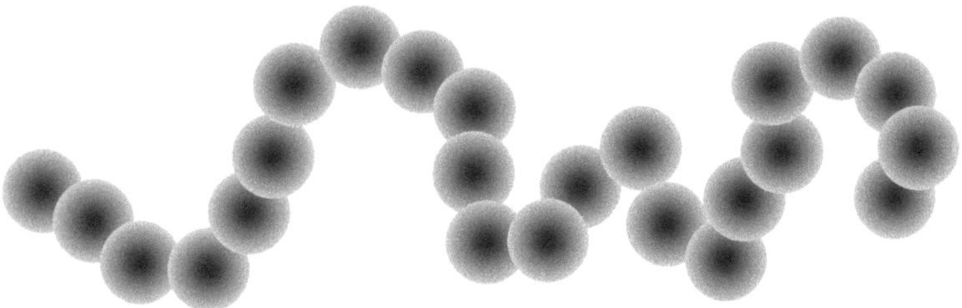

Pearl-necklace model is very interesting to look at. Pearls are the monomer units of the polymer.

viii. Lattice chain model of polymer

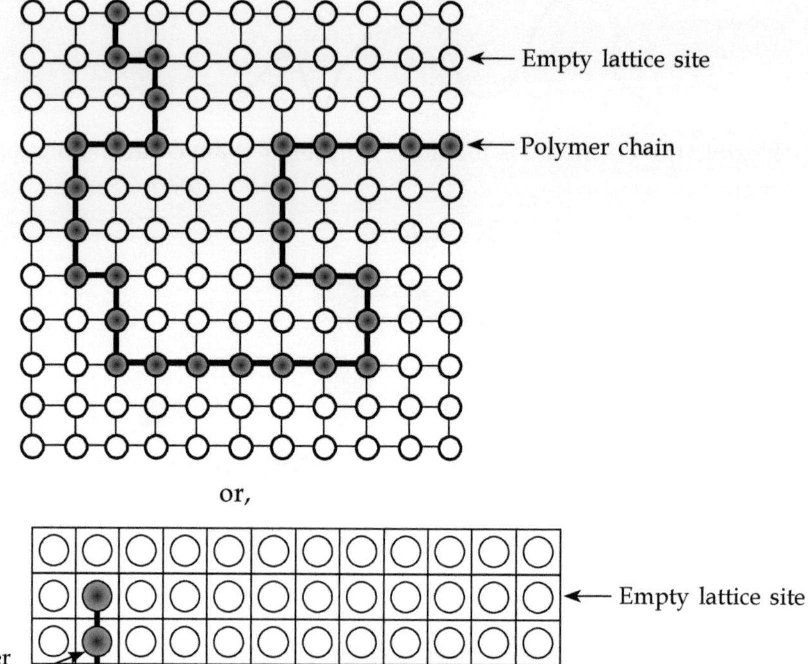

or,

8.5 Flory–Huggins Theory

From the thermodynamic point of view, dissolution of a polymer into a solvent is favourable when Gibbs free energy of mixing is negative. This can happen when either ΔH_{mix} is negative or the product ($T\Delta S_{\text{mix}}$) of the temperature and the entropy of mixing is greater than the enthalpy of mixing (ΔH_{mix}). The amount of change in entropy of mixing (ΔS_{mix}) of polymer solution is always positive and relatively smaller than small molecules. Anyhow you have to calculate the values of ΔH_{mix} and ΔS_{mix} for the determination of ΔG_{mix}. The Flory–Huggins theory uses the lattice model to arrange the polymer chains and solvent molecules for the determination of ΔH_{mix} and ΔS_{mix}. The simplest version of this lattice chain theory is generally referred to as Flory–Huggins mean-field theory. It is assumed that there is no volume change for polymer solution mixing ($\Delta V_{\text{mix}} = 0$). Thus the expression of ΔS_{mix} for polymer in solution will be (using the Flory-Huggins theory)

$$\Delta S_{\text{mix}} = -R\left[\phi_1 \ln\phi_1 + \frac{\phi_2}{n}\ln\phi_2\right]$$

where ϕ_1, ϕ_2 are the volume fractions or number fractions of the component 1 and 2. n is the lattice sites occupied per polymer or degree of polymerization.

The expression of ΔH_{mix} for polymer in solution will be

$$\Delta H_{mix} = RT\chi_{12}\phi_1\phi_2$$

where χ_{12} is the Flory-Huggins binary interaction parameter, R is the universal gas constant and T is the absolute temperature.

Thus the expression of ΔG_{mix} for polymer in solution will be

$$\Delta G_{mix} = RT\left[\phi_1\ln\phi_1 + \frac{\phi_2}{n}\ln\phi_2 + \chi_{12}\,\phi_1\phi_2\right]$$

 ## Entropy of polymer mixing

Now consider step by step to determine the entropy of polymer mixing. For simplicity first consider regular solutions and then will move to polymer solutions. The Flory–Huggins theory uses the lattice model to arrange the two components in lattice sites. When two or more components (chemical species) are mixed together mixtures are formed. In case of two components mixing it is termed as binary mixtures, for three components it is termed as ternary mixtures. Whatever be the situation entropy always favours mixing. See the schematic presentation below to understand the entropy of mixing.

Consider two components having volume V_1 and V_2 respectively and the total volume V ($V = V_1 + V_2$).

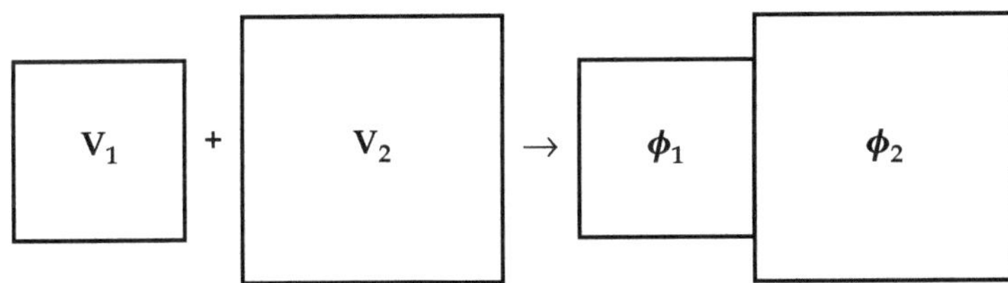

Thus volume fraction of component 1 will be

$$\phi_1 = \frac{V_1}{V_1 + V_2} = \frac{V_1}{V}$$

Similarly volume fraction of component 2 will be

$$\phi_2 = \frac{V_2}{V_1 + V_2} = \frac{V_2}{V}$$

[For better understanding take the help of an example. Suppose two components having volume 10 L and 40 L respectively. Total volume (10 + 40 = 50 L). Then volume fraction of component 1 is $\frac{10}{50} = 0.2$ and the volume fraction of component 2 is 0.8.]

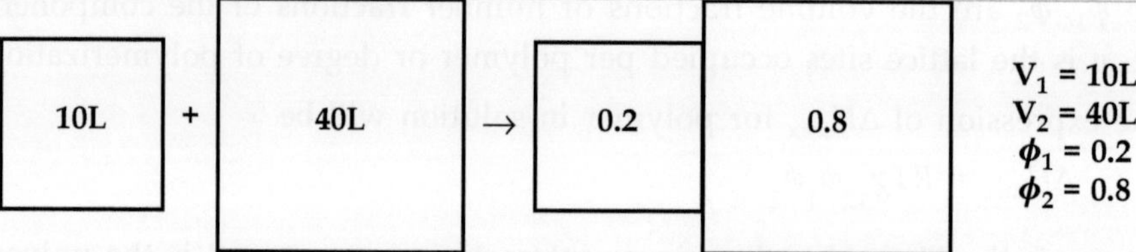

Now consider the following schematic presentation.

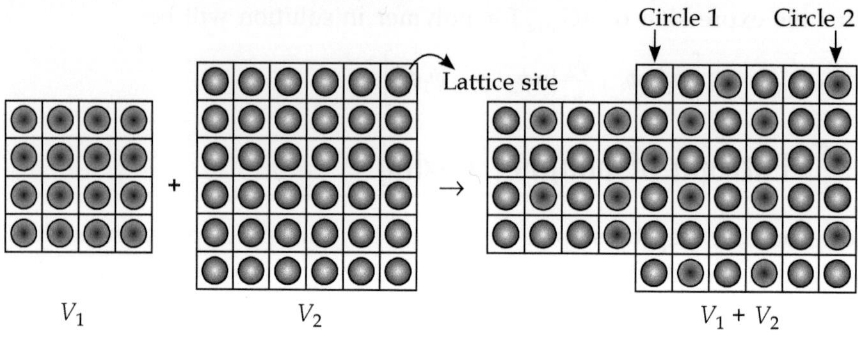

The volumes of two components are still V_1 and V_2 but V_1 & V_2 are made by n_1 & n_2 number of lattice sites. Again the volume of each sites is v_0 and total sites are n $(n_1 + n_2 = n)$. Thus $V_1 = n_1 v_0$, $V_2 = n_2 v_0$ and $V = n v_0$. This is an assumption that components 1 and 2 both have the same volume of their lattice site. Clearly the volume of circle 1 and circle 2 is equal and that is v_0. Keep in mind that it is a 2D presentation, in 3D presentation circle will be sphere. Thus volume fraction can be replaced by number fraction.

Thus number fraction of component 1 will be

$$\phi_1 = \frac{V_1}{V_1 + V_2} = \frac{V_1}{V} = \frac{n_1 v_0}{n v_0} = \frac{n_1}{n}$$

and the number fraction of component 2 will be

$$\phi_2 = \frac{V_2}{V_1 + V_2} = \frac{V_2}{V} = 1 - \phi_1 = \frac{n_2}{n}$$

Now consider the polymer solution which means polymer mixed with solvent. See the following schematic presentation.

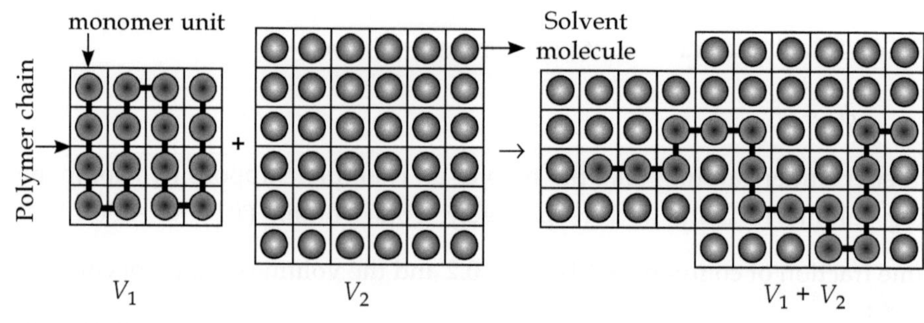

A polymer chain is formed by combining many lattice sites which depends on the monomer units. Each monomer unit occupies one lattice site. Let one polymer chain is formed by combining n_m number of monomer units and there are n_p number of polymer chains. But each solvent molecule occupies one lattice site. ϕ_p is the volume fraction of polymer and n is the total sites. Thus,

Number of polymer chains = total lattice sites ×

$$\frac{\text{volume fraction of polymer}}{\text{lattice sites occupied by single polymer chain}}$$

i.e., $n_p = n\dfrac{\phi_p}{n_m}$ or, $\phi_p = \dfrac{n_p n_m}{n}$

- It will be easier to understand with mathematical examples. Let mix polymer [polymer contain 10 (n_p) polymer chains, each chain has 200 (n_m) monomer units. Thus it will occupy 2000 lattice sites] with solvent molecules (occupy 8000 lattice sites). Total lattice site n = 10000, then $10 = 10000 \times \dfrac{0.2}{200}$ as

$$\phi_p(0.2) = \frac{2000}{10000}.$$

And the number of solvent molecules is $n_s = n(1 - \phi_p)$ or, $\phi_s = \dfrac{n_s}{n}$

where n_s and ϕ_s are the number of lattice sites occupied by solvent molecules and volume fraction of solvent respectively.

Once we have number of polymer chains and solvent molecules, entropy can be calculated using the equation

$$S = k\ln\Omega$$

where S is entropy, k is Boltzmann constant and Ω is the arrangement (possible distribution or number of states) of polymer chains and solvent molecules on the lattice. These two equations are needed to calculate S

$$\Omega = \frac{N!}{n_p!\,n_s!} \text{ and Stirling's approximation } \ln x! \approx x\ln x - x$$

There are four states, entropy of polymer before mixing and after mixing, entropy of solvent before mixing and after mixing. Therefore, we can write down the entropy per molecule as

S_p^B = entropy of one polymer chain before mixing = $k\ln(n_p n_m v_0)$

S_s^B = entropy of one solvent molecule before mixing
   $= k\ln(n_s v_0)$, as each solvent molecule occupy one site

S_p^A = entropy of one polymer chain after mixing = $k\ln[n_p n_m + n_s)v_0]$

S_s^A = entropy of one solvent molecule after mixing = $k\ln[n_p n_m + n_s)v_0]$

Thus total entropy change due to polymer mixing with solvents

$$\Delta S_{\text{mix}} = n_p S_p^A + n_s S_s^A - n_p S_p^B - n_s S_s^B$$

$$= -k\left[n_p \ln\left(\frac{n_p n_m}{n_p n_m + n_s}\right) + n_s \ln\left(\frac{n_s}{n_p n_m + n_s}\right)\right] = -k[n_p \ln\phi_p + n_s \ln\phi_s]$$

Divided by n, the equation becomes

$$\frac{\Delta S_{\text{mix}}}{n} = -k\left[\frac{n_p}{n}\ln\phi_p + \frac{n_s}{n}\ln\phi_s\right] = -k\left[\frac{\phi_p}{n_m}\ln\phi_p + \phi_s\ln\phi_s\right]$$

 Enthalpy of polymer mixing

Enthalpy of polymer mixing can be calculated by following steps (i) counting total number of interactions of polymer–solvent system before and after mixing (ii) values of their interaction energies.

Let u_{ss}, u_{pp} and u_{ps} are the interaction energies between adjacent lattice sites for a solvent-solvent (S—S) contact, a polymer-polymer (P—P) contact, and a polymer-solvent (P—S) contact respectively. Each lattice site has a definite coordination number (Z) depending on the lattice. For square lattice the coordination number $Z = 4$ and for cubic lattice the coordination number $Z = 6$ respectively. Z also denotes the number of adjacent lattice sites.

This is a square lattice

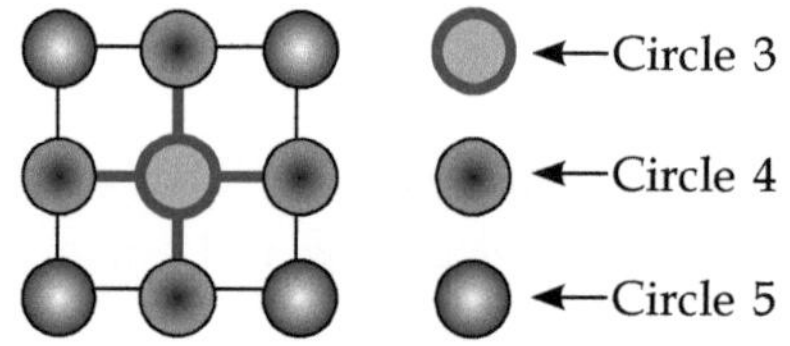

Each lattice site (**Circle 3**) has 4 interactions (shown as **bold line**) with adjacent lattice sites (shown as **Circle 4**). There are no interactions if they are not adjacent (shown as **Circle 5**).

The square lattice of polymer-solvent interaction can be shown as

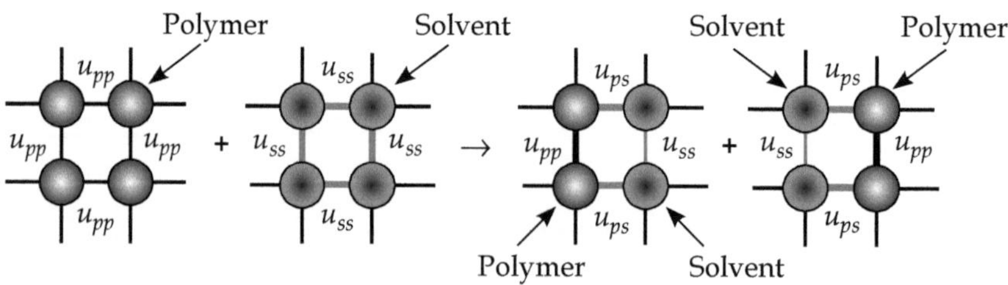

Here total sites are eight (four for polymer and four for solvents). Before mixing there are four P—P contacts and four S—S contacts and after mixing there are two P—P contacts, two S—S contacts and four P—S contacts. Let the interaction energies of P—P contact, S—S contact and P—S contact are u_{pp}, u_{ss} and u_{ps} respectively.

The interaction energy before mixing $= 4u_{pp} + 4u_{ss}$

The interaction energy after mixing $= 2u_{pp} + 2u_{ss} + 4u_{ps}$

Total energy change $= 2u_{pp} + 2u_{ss} + 4u_{ps} - (4u_{pp} + 4u_{ss}) = 4u_{ps} - 2(u_{pp} + u_{ss})$

There are **eight** sites or eight bonds from which **two** P—S contacts or bonds formed. Thus for each newly created P—S contact, the energy change is

$$\frac{\text{two}}{\text{eight}}\,[4u_{ps} - 2(u_{pp} + u_{ss})] = \frac{2}{8}\,[4u_{ps} - 2(u_{pp} + u_{ss})] = u_{ps} - \frac{(u_{pp} + u_{ss})}{2}$$

The χ(chi) parameter, also called Flory's χ parameter or Flory–Huggins χ parameter, is expressed as

$$\chi = \frac{Z\left[u_{ps} - \dfrac{(u_{pp} + u_{ss})}{2}\right]}{kT}$$

where Z is coordination number, k is Boltzmann constant and T is absolute temperature.

When χ = (-)ve, P—S contacts are favoured and for χ = (+)ve, P—P & S—S contacts are favoured.

Probability of adjacent contacts are

contact	Interaction energies	Probability before mixing	Probability after mixing
P—P	u_{pp}	ϕ_p	$(\phi_p)^2$
S—S	u_{ss}	$\phi_s = 1 - \phi_p$	$(\phi_s)^2 = (1 - \phi_p)^2$
P—S	u_{ps}	0	$2\phi_p\phi_s = 2\phi_p(1 - \phi_p)$

Now we will count the total number of interactions of polymer-solvent system before and after mixing with the help of probability. The probability for formation of a P—P contact, S—S contact and P—S contact are $(\phi_p)^2$, $(\phi_s)^2$ or $(1 - \phi_p)^2$ and $2\phi_p\phi_s$ or $2\phi_p(1 - \phi_p)$ respectively.

Total interaction energy before mixing

For polymer $= \dfrac{Zn}{2}\,u_{pp}\phi_p$

For solvent $= \dfrac{Zn}{2}\,u_{ss}\phi_s = \dfrac{Zn}{2}\,u_{ss}(1 - \phi_p)$

For both polymer and solvent $= \dfrac{Zn}{2}\,u_{pp}\phi_p + \dfrac{Zn}{2}\,u_{ss}(1 - \phi_p)$

$$= \frac{Zn}{2}\,[u_{pp}\phi_p + u_{ss}(1 - \phi_p)] = H_B$$

And total interaction energy after mixing

$$H_A = \frac{Zn}{2}\,[\{u_{pp}\phi_p + u_{ps}(1 - \phi_p)\}\phi_p + \{u_{ps}\phi_p + u_{ss}(1 - \phi_p)\}(1 - \phi_p)]$$

$$= \frac{Zn}{2}\,[u_{pp}(\phi_p)^2 + 2u_{ps}\phi_p(1 - \phi_p) + u_{ss}(1 - \phi_p)^2]$$

Thus, total interaction energy for mixing is

$$\Delta H_{mix} = H_A - H_B$$

$$= \frac{Zn}{2}\,[u_{pp}(\phi_p)^2 + 2u_{ps}\phi_p(1 - \phi_p) + u_{ss}(1 - \phi_p)^2] - \frac{Zn}{2}\,[u_{pp}\phi_p + u_{ss}(1 - \phi_p)]$$

$$= \frac{Zn}{2}[u_{pp}(\phi_p)^2 + 2u_{ps}\phi_p(1 - \phi_p) + u_{ss}(1 - \phi_p)^2 - u_{pp}\phi_p - u_{ss}(1 - \phi_p)]$$

$$= \frac{Zn}{2}[u_{pp}\{(\phi_p)^2 - \phi_p\} + 2u_{ps}\phi_p(1 - \phi_p) + u_{ss}\{1 - 2\phi_p + (\phi_p)^2 - 1 + \phi_p\}]$$

$$= \frac{Zn}{2}[u_{pp}\phi_p(\phi_p - 1) + 2u_{ps}\phi_p(1 - \phi_p) + u_{ss}\phi_p(\phi_p - 1)]$$

$$= \frac{Zn}{2}[-u_{pp}\phi_p(1 - \phi_p) + 2u_{ps}\phi_p(1 - \phi_p) - u_{ss}\phi_p(1 - \phi_p)]$$

$$= \frac{Zn}{2}\phi_p(1 - \phi_p)(2u_{ps} - u_{pp} - u_{ss})$$

Hence interaction energy for mixing per lattice site is

$$= (\Delta H_{mix}/n) = [\frac{Zn}{2}\phi_p(1 - \phi_p)(2u_{ps} - u_{pp} - u_{ss})]/n$$

$$= \frac{Z}{2}\phi_p(1 - \phi_p)(2u_{ps} - u_{pp} - u_{ss}) = \chi\phi_p(1 - \phi_p)kT$$

where $\dfrac{Z\left[u_{ps} - \dfrac{(u_{pp} + u_{ss})}{2}\right]}{kT}$

χ (chi) is Flory's χ parameter or Flory–Huggins χ parameter.

It was assumed in Flory–Huggins theory that there is no volume change after mixing, again the monomer units of polymer chain and solvent molecule have fit on the same lattice sites. But in reality, volume changes after mixing and also packing mismatch in lattice sites. These volume effects lead to deviation from lattice site model and sum up into the interaction parameter χ which have nontrivial dependency on composition, chain length and temperature. Thus $\chi(T)$ can be written as

$$\chi(T) \cong A + \frac{B}{T}$$

where A is temperature independent entropic part and $\dfrac{B}{T}$ is enthalpic part.

The expression of Flory-Huggins equation for polymer in solution is

$$\Delta G_{mix} = \Delta H_{mix} - T\Delta S_{mix}$$

$$= kT\chi\phi_p(1 - \phi_p) - T\left[-k\left(\frac{\phi_p}{n_m}\ln\phi_p + \phi_s\ln\phi_s\right)\right]$$

$$= kT\left[\frac{\phi_p}{n_m}\ln\phi_p + \phi_s\ln\phi_s + \chi\phi_p(1 - \phi_p)\right]$$

Flory-Huggins equation of solutions has been used to study the phase-separation phenomena in polymer solutions. The drawback of this theory is that, the interaction parameter is assumed to be independent of mixture composition. It will not work when the composition-dependent interaction parameter vary with tempearture. It will also not work for hollow-fiber, hollow-sphere polymers.

8.6 Lower and Upper Critical Solution Temperatures

There are varieties of Flory Huggins phase diagram depending on the nature of solute and solvent molecules. In the solute-solvent systems, solute can be small molecule or polymer and solvent can be liquid molecule or another polymer. Thus the systems are small molecule-solvent, polymer-solvent (polymer solution) and polymer-polymer (blend) system. Though polymer solutions phase diagram is of main interest but other systems will also be shown. Flory–Huggins interaction parameter $\chi(T)$ can be written as

$$\chi(T) \cong A + \frac{B}{T}$$

When B > 0, with increasing temperature χ decreases and the highest temperature of the two-phase region is termed as upper critical solution temperature (UCST), T_c. When the solution temperature is more than T_c, it makes a stable homogeneous solution.

When B < 0, with increasing temperature χ increases and the lowest temperature of two-phase region is termed as lower critical solution temperature (LCST), T_c. When the solution temperature is less than T_c, it makes a stable homogeneous solution.

The schematic presentations of different Flory Huggins phase diagrams are shown below.

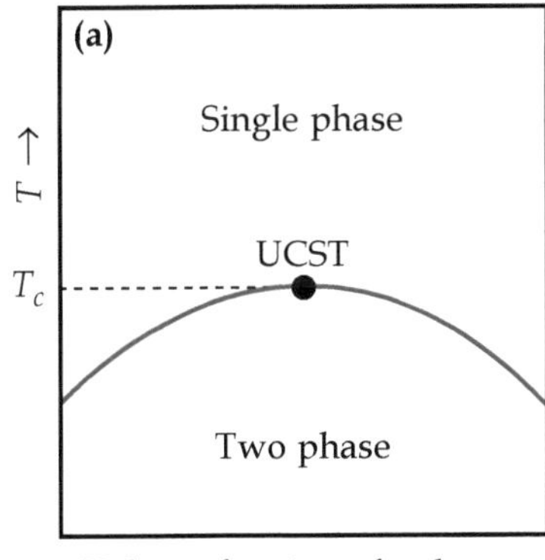

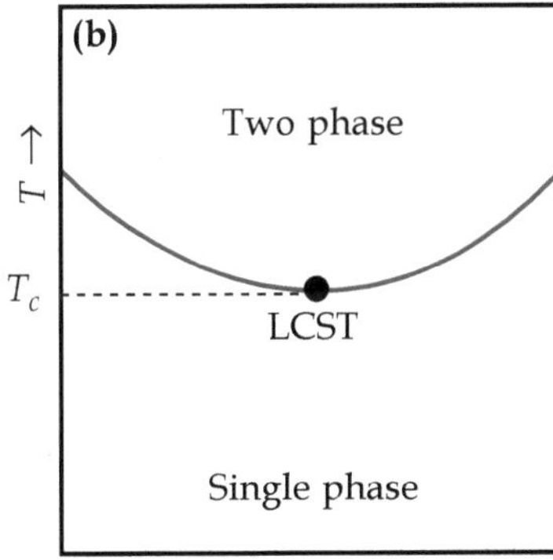

(a) This phase diagram is for those polymers/mixtures which have only UCST. Above the UCST, it remains as single phase.

(b) This phase diagram is for those polymers/mixtures which have only LCST, below the LCST it remains as single phase.

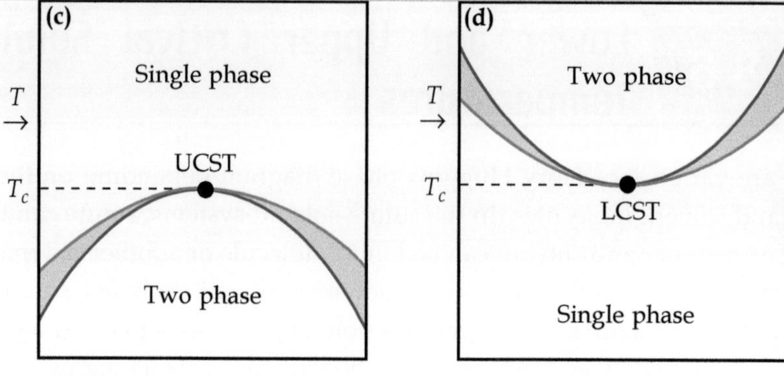

(c) This phase diagram explain the UCST and metastable region of the polymers/mixtures.

(d) This phase diagram explain the LCST and metastable region of the polymers/mixtures.

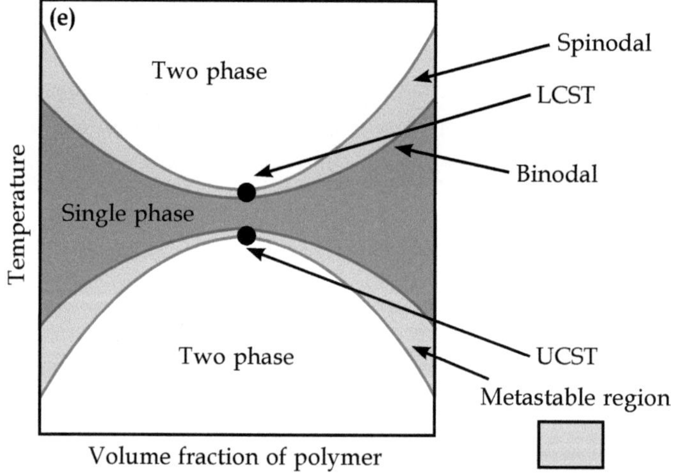

(e) This phase diagram has low miscibility region. The solution remains in the single phase between two binodal curves. It has both UCST and LCST, two two-phase regions and one single-phase region.

1. **What do you understand about solubility parameter?**

Ans. The numerical value for the square root of the cohesive energy density which indicates thesolvency behaviour of a specific solvent called Hildebrand solubility parameter or solubility parameter. The solubility parameter δ_i of substance i is mathematically expressed as

$$\delta_i = \left(c_i\right)^{\frac{1}{2}} = \left[\frac{\Delta H_i - RT}{V_i}\right]^{\frac{1}{2}}$$

where c_i, ΔH_i and V_i are the cohesive energy density, heat of vaporization and molar volume of substance i, R and T are gas constant and temperature respectively.

2. What do you understand by polymer solubility?

Ans. Solubility of polymers mainly depends on nature of polymer (size, shape and dimension of polymer molecules), molecular weight etc. of polymer as well as nature of solvent molecule and temperature of the polymeric solution. Smaller the dimension of polymers, greater will be the polymer - solvent interaction, higher will be the chances of polymers becoming soluble. Thus lower molecular weight polymers have more tendency to come into solution (more soluble) than high molecular weight polymers. The interaction mainly will depend on van der Waals forces, hydrogen bonding, polarity, heat ofvaporization and cohesive energy density etc. acting on the total system. It is a guessing phenomenon that polar solvent dissolves polar polymers and non-polar solvent dissolves non-polar polymers.In thermodynamic point of view, energy required when a polymer will comes into solution. The energy used will be in the following form namely mechanical shaking, stirring, sonication, vortexing and others form. The solvent molecules penetrate into polymer network structure or inserted into polymer lattice structure or swell the bulk polymer. For this step energy is required and the amount of energy will be higher for ordered structure polymers. Thus dissolving a rubbery polymer will be easier than dissolving a crystalline polymer in terms of energy requirement. When temperature is increased, energy is also increased and thus polymers have higher solubility. But few polymers also show reverse effects.

3. Explain shortly the thermodynamics of polymer solutions.

Ans. When a polymer is added in a solvent, the state of the pure solvent changes to another state. The change of G, H and S of these two states are ΔG (change in Gibbs free energy), ΔH (change in enthalpy) and ΔS (change in entropy) respectively. Thus Gibbs free-energy for polymer mixing are expressed as

$$\Delta G_{mix} = \Delta H_{mix} - T\Delta S_{mix}$$

A negative ΔG means that the reactantshave more free energy than the products. As the term $T\Delta S_{mix}$ is always negative, the sign of ΔG_{mix} will mainly be determined by ΔH_{mix} (its sign and magnitude). A negative value of ΔG_{mix} indicates that a solvent-polymer system forms a homogeneous solution (polymer and solvent aremiscible).

4. Calculate the entropy of polymer chains and solvent molecules before and after mixing.

Ans. The entropy of polymer chains and solvent molecules before and after mixing can be calculated using the equation

$$S = k \ln \Omega$$

where S is entropy, k is Boltzmann constant and Ω is the arrangement (possible distribution or number of states) of polymer chains and solvent molecules on the lattice. These two equations needed to calculate S.

$$\Omega = \frac{N!}{n_p! \, n_s!} \quad \text{and Stirling's approximation } \ln x! \approx x \ln x - x$$

There are four states, entropy of polymer before mixing and after mixing, entropy of solvent before mixing and after mixing. Therefore, we can write down the entropy per molecule as

S_p^B = entropy of one polymer chain before mixing = $k \ln (n_p n_m v_0)$

S_s^B = entropy of one solvent molecule before mixing
$$= k \ln (n_s v_0), \text{ as each solvent molecule occupy one site}$$

S_p^A = entropy of one polymer chain after mixing = $k \ln[n_p n_m + n_s)v_0]$

S_s^A = entropy of one solvent molecule after mixing = $k \ln[n_p n_m + n_s)v_0]$

5. Derive enthalpy of polymer mixing.

Ans. Enthalpy of polymer mixing can be calculated by following steps (i) counting total number of interactions of polymer–solvent system before and after mixing (ii) values of their interaction energies.

Let u_{ss}, u_{pp}, and u_{ps} are the interaction energies between adjacent lattice sites for a solvent–solvent (S—S) contact, a polymer–polymer (P—P) contact, and a polymer–solvent (P—S) contact respectively. Each lattice site has a definite coordination number (Z) depending on the lattice. For square lattice the coordination number $Z = 4$ and for cubic lattice the coordination number $Z = 6$ respectively. Z also denotes the number of adjacent lattice sites.

This is a square lattice

Each lattice site (**Circle 3**) has 4 interactions (shown as **bold line**) with adjacent lattice sites (shown as **Circle 4**). There are no interactions if they are not adjacent (shown as **Circle 5**).

The square lattice of polymer-solvent interaction can be shown as

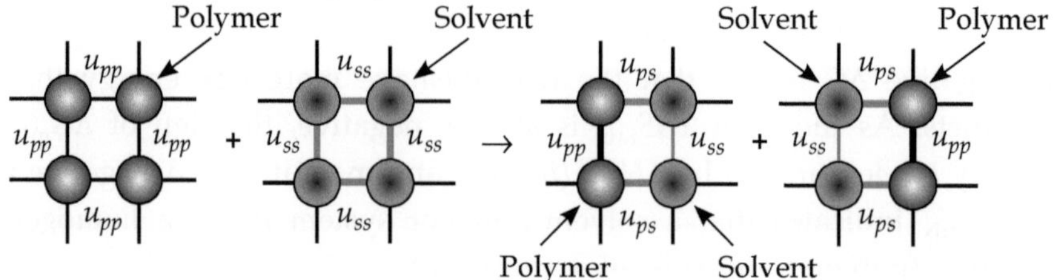

Here total sites are eight (four for polymer and four for solvents). Before mixing there are four P—P contacts and four S—S contacts and after mixing there are two P—P contacts, two S—S contacts and four P—S contacts. Let the interaction energies of P—P contact, S—S contact and P—S contact are u_{pp}, u_{ss} and u_{ps} respectively.

The interaction energy before mixing $= 4u_{pp} + 4u_{ss}$

The interaction energy after mixing $= 2u_{pp} + 2u_{ss} + 4u_{ps}$

Total energy change $= 2u_{pp} + 2u_{ss} + 4u_{ps} - (4u_{pp} + 4u_{ss}) = 4u_{ps} - 2(u_{pp} + u_{ss})$

There are eight sites or eight bonds from which two P—S contacts or bonds formed. Thus for each newly created P—S contact, the energy change is

$$\frac{\text{two}}{\text{eight}} [4u_{ps} - 2(u_{pp} + u_{ss})] = \frac{2}{8} [4u_{ps} - 2(u_{pp} + u_{ss})] = u_{ps} - \frac{(u_{pp} + u_{ss})}{2}$$

The χ (chi) parameter, also called Flory's χ parameter or Flory–Huggins χ parameter, is expressed as

$$\chi = \frac{Z\left[u_{ps} - \frac{(u_{pp} + u_{ss})}{2}\right]}{kT}$$

where Z is coordination number, k is Boltzmann constant and T is absolute temperature. When $\chi = (-)$ve, P—S contacts are favoured and for $\chi = (+)$ve, P—P & S—S contacts are favoured.

Probability of adjacent contacts are

Contact	Interaction energies	Probability before mixing	Probability after mixing
P—P	u_{pp}	ϕ_p	$(\phi_p)^2$
S—S	u_{ss}	$\phi_s = 1 - \phi_p$	$(\phi_s)^2 = (1 - \phi_p)^2$
P—S	u_{ps}	0	$2\phi_p\phi_s = 2\phi_p(1 - \phi_p)$

Now we will count the total number of interactions of polymer–solvent system before and after mixing with the help of probability. The probability for formation of a P—P contact, S—S contact and P—S contact are $(\phi_p)^2$, $(\phi_s)^2$ or $(1 - \phi_p)^2$ and $2\phi_p\phi_s$ or $2\phi_p(1 - \phi_p)$ respectively.

Total interaction energy before mixing

For polymer $= \dfrac{Zn}{2} u_{pp}\phi_p$

For solvent $= \dfrac{Zn}{2} u_{ss}\phi_s = \dfrac{Zn}{2} u_{ss}(1 - \phi_p)$

For both polymer and solvent $= \dfrac{Zn}{2} u_{pp}\phi_p + \dfrac{Zn}{2} u_{ss}(1 - \phi_p)$

$$= \frac{Zn}{2} [u_{pp}\phi_p + u_{ss}(1 - \phi_p)] = H_B$$

And total interaction energy after mixing

$$HA = \frac{Zn}{2} [\{u_{pp}\phi_p + u_{ps}(1 - \phi_p)\}\phi_p + \{u_{ps}\phi_p + u_{ss}(1 - \phi_p)\}(1 - \phi_p)]$$

$$= \frac{Zn}{2} [u_{pp}(\phi_p)^2 + 2u_{ps}\phi_p(1 - \phi_p) + u_{ss}(1 - \phi_p)^2]$$

Thus, total interaction energy for mixing is

$$\Delta H_{mix} = H_A - H_B$$

$$= \frac{Zn}{2}[u_{pp}(\phi_p)^2 + 2u_{ps}\phi_p(1 - \phi_p) + u_{ss}(1 - \phi_p)^2] - \frac{Zn}{2}[u_{pp}\phi_p + u_{ss}(1 - \phi_p)]$$

$$= \frac{Zn}{2}[u_{pp}(\phi_p)^2 + 2u_{ps}\phi_p(1 - \phi_p) + u_{ss}(1 - \phi_p)^2 - u_{pp}\phi_p - u_{ss}(1 - \phi_p)]$$

$$= \frac{Zn}{2}[u_{pp}\{(\phi_p)^2 - \phi_p\} + 2u_{ps}\phi_p(1 - \phi_p) + u_{ss}\{1 - 2\phi_p + (\phi_p)^2 - 1 + \phi_p\}]$$

$$= \frac{Zn}{2}[u_{pp}\phi_p(\phi_p - 1) + 2u_{ps}\phi_p(1 - \phi_p) + u_{ss}\phi_p(\phi_p - 1)]$$

$$= \frac{Zn}{2}[-u_{pp}\phi_p(1 - \phi_p) + 2u_{ps}\phi_p(1 - \phi_p) - u_{ss}\phi_p(1 - \phi_p)]$$

$$= \frac{Zn}{2}\phi_p(1 - \phi_p)(2u_{ps} - u_{pp} - u_{ss})$$

Hence interaction energy for mixing per lattice site is

$$= (\Delta H_{mix}/n) = [\frac{Zn}{2}\phi_p(1 - \phi_p)(2u_{ps} - u_{pp} - u_{ss})]/n$$

$$= (Z/2)\phi_p(1 - \phi_p)(2u_{ps} - u_{pp} - u_{ss}) = \chi\phi_p(1 - \phi_p)kT$$

Where $\chi = \dfrac{Z\left[u_{ps} - \dfrac{(u_{pp} + u_{ss})}{2}\right]}{kT}$

χ (chi) is Flory's χ parameter or Flory–Huggins χ parameter.

6. **Write down the expression of Flory–Huggins equation for polymer in solution.**

Ans. The expression of Flory–Huggins equation for polymer in solution is

$$\Delta G_{mix} = \Delta H_{mix} - T\Delta S_{mix}$$

$$= kT\chi\phi_p(1 - \phi_p) - T\left[-k\left(\frac{\phi_p}{n_m}\ln\phi_p + \phi_s\ln\phi_s\right)\right]$$

$$= kT\left[\frac{\phi_p}{n_m}\ln\phi_p + \phi_s\ln\phi_s + \chi\phi_p(1 - \phi_p)\right]$$

7. **Show schematically lower and upper critical solution temperatures.**

Ans. Few Flory Huggins phase diagrams are shown below

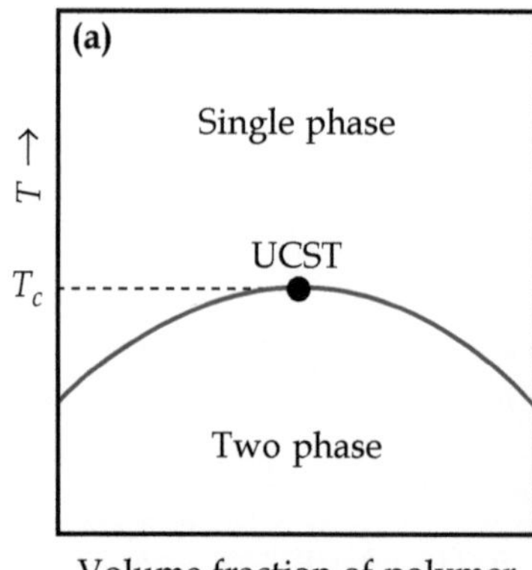

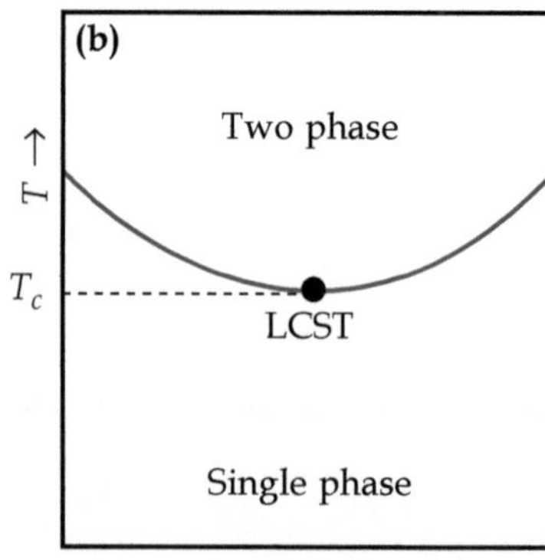

(a) This phase diagram is for those polymers/mixtures which have only UCST. Above the UCST, it remains as single phase.

(b) This phase diagram is for those polymers/mixtures which have only LCST below the LCST it remains as single phase.

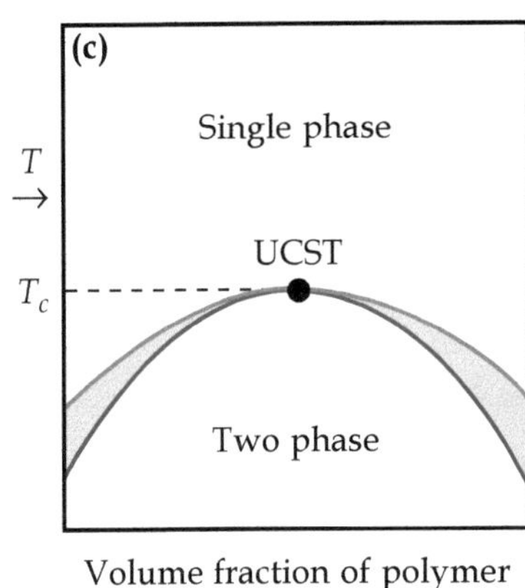

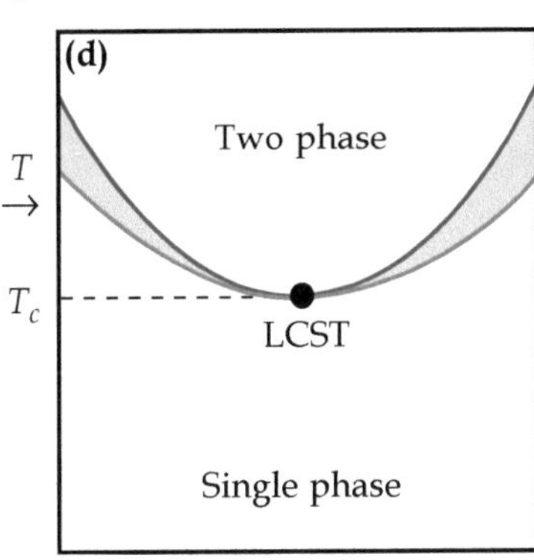

(c) This phase diagram explain the UCST and metastable region of the polymers/mixtures.

(d) This phase diagram explain the LCST and metastable region of the polymers/mixtures.

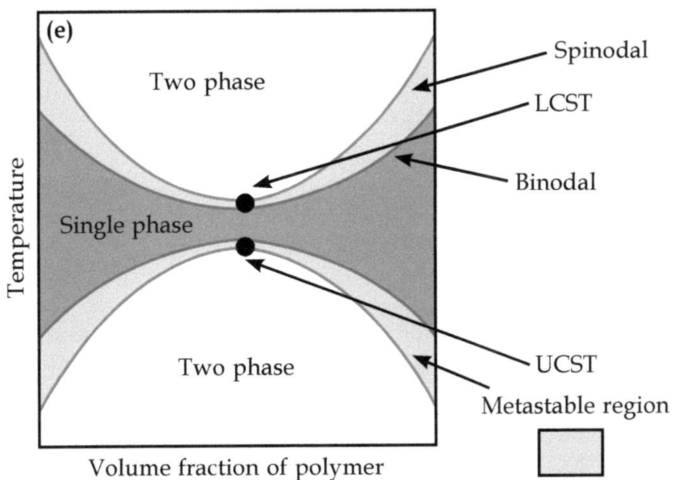

(e) This phase diagram has low miscibility region. The solution remains in the single phase between two binodal curves. It has both UCST and LCST, two two-phase regions and one single-phase region.

Flory–Huggins interaction parameter $\chi(T)$ can be written as

$$\chi(T) \cong A + \frac{B}{T}$$

When B > 0, with increasing temperature χ decreases and the highest temperature of the two-phase region is termed as upper critical solution temperature (UCST), T_c. When the solution temperature is more than T_c, it makes a stable homogeneous solution.

When B < 0, with increasing temperature χ increases and the lowest temperature of two-phase region is termed as lower critical solution temperature (LCST), T_c. When the solution temperature is less than T_c, it makes a stable homogeneous solution.

Exercise

A Multiple Choice Type Questions(MCQ) (Each Question carries 1 mark)

1. What is true of the enthalpy of mixing, ΔH_m, for an ideal solution?

   a. $\Delta H_m > 0$ **b.** $\Delta H_m = 0$

   c. $\Delta H_m < 0$ **d.** None of these

2. What is true of the entropy of mixing, ΔS_m, for an ideal solution?

   a. $\Delta S_m < 0$ **b.** $\Delta S_m = 0$

   c. $\Delta S_m > 0$ **d.** None of these

3. What is true of the Gibbs free energy of mixing, ΔG_m, for an ideal solution?

   a. $\Delta G_m < 0$ **b.** $\Delta G_m = 0$

   c. $\Delta G_m > 0$ **d.** None of these

4. Which is more favourable for mixing

   a. low Flory-Huggins parameter

   b. high Flory-Huggins parameter

   c. independent of Flory-Huggins parameter

   d. none of these

5. Which of these polymers would be a polyelectrolyte when dissolved in water?

   a. polyacrylic acid

   b. poly (methyl acrylate)

   c. sodium polyacroylate

   d. sodium polyacroylate andpolyacrylic acid

6. If you added NaCl to an aqueous solution of sodium polyacrylate, how would the polymer conformation change?

   a. The polymer would become more coiled (shrink)

   b. Nothing would change

   c. The polymer would be more extended (swell)

   d. Explosion occurs

7. The entropy change of mixing for an ideal solution formed by mixing 1.9 mole of the solventwith 0.1 mole of the solute at 300K is

a. 1.3 J/K　　　　　　　　　　　　**b.** 4.3 J/K

c. 2.3 J/K　　　　　　　　　　　　**d.** 3.3 J/K

8. A polymer solution is formed by mixing 1.9 mol of the solvent and 0.1 mol of the polymer containing 100 segments per polymer molecule. If the Flory-Huggins polymer-solvent interaction parameter is 0.3, the Gibbsfree energy change of mixing associated with the formation of the polymersolution at 300 K is

a. – 3142 J　　　　　　　　　　　　**b.** – 1542 J

c. – 7542 J　　　　　　　　　　　　**d.** – 5942 J

9. Flory–Huggins χ parameter, is expressed as (the terms are usual)

a. $\chi = \dfrac{Zu_{ps} - \dfrac{\left(u_{pp}+u_{ss}\right)}{2}}{kT}$　　　　　　**b.** $\chi = \dfrac{Zu_{ps} - \dfrac{\left(u_{pp}+u_{ss}\right)}{2}}{T}$

c. $\chi = \dfrac{Zu_{ps} - \dfrac{\left(u_{pp}+u_{ss}\right)}{2}}{k}$　　　　　　**d.** $\chi = \dfrac{Zu_{ps} - \dfrac{\left(u_{pp}+u_{ss}\right)}{2}}{2R}$

10. Flory–Huggins interaction parameter $\chi(T)$ can be written as

$$\chi(T) \cong A + \frac{B}{T}$$

When B > 0, with increasing temperature χ

a. decreases　　　　　　　　　　　**b.** increases

c. independent　　　　　　　　　　**d.** none of these

B **Short Answer Type Questions (Each question carry either 2 or 3 marks)**

1. What are the different criteria of polymer solubility?

2. Why polymers need to bring into solution?

3. What are ideal solubility parameters?

4. How do you measure solubility parameters?

5. What are polymer additives?

6. Which polymer additives are added to improve flexibility?

7. What is the change of entropy associated with an ideal chain of N monomersforming a ring?

8. Show schematically lower and upper critical solution temperatures

C **Long Answer Type Questions (Each question carry 5 marks)**

1. Describe the thermodynamics of an ideal chain stretching.

2. Describe the thermodynamics of an ideal chain squashing.

3. An ideal polymer chain is end-tethered to a fixed wall and a weight of massm is suspended from the free-end of the chain, as pictured below. The weight issufficiently small that the elongation of the chain, R, is between the natural sizeof the chain and its contour length, $\sqrt{(N)}_a < R < N_a$. Derive an expression for thechain elongation, x, as a function of the mass of the weight, m, under isothermalconditions. (critical)

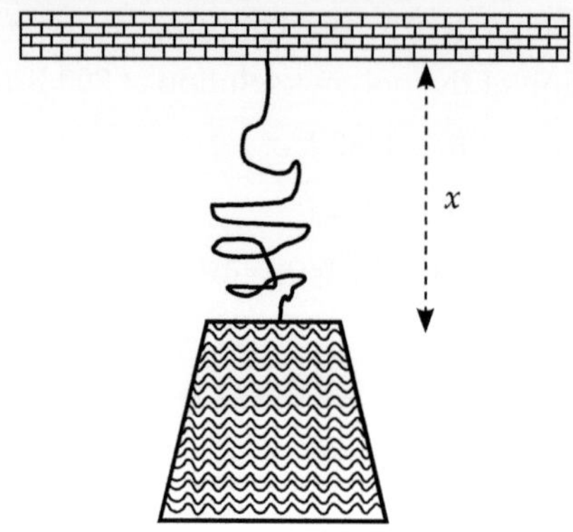

[**Hints :** $F = U - TS$

$$f = -\frac{dF}{dR} = T\frac{dS}{dR}$$

$$S(R) = k_B \ln [P(R)]$$

$$\frac{dS}{dR} = k_B \frac{d}{dR}\ln\left[\left(\frac{3}{2\pi Na^2}\right)^{\frac{3}{2}}\exp\left(-\frac{3R^2}{2Na^2}\right)\right]$$

$$\frac{dS}{dR} = k_B \frac{d}{dR}\ln\left[\left(\frac{3}{2\pi Na^2}\right)^{\frac{3}{2}}\right] + k_B\frac{d}{dR} + k_B\frac{d}{dR}\left(-\frac{3R^2}{2Na^2}\right)$$

Neglect 1st term $\dfrac{dS}{dR} = -k_B\dfrac{3R}{Na^2}$

Hence $f = -k_B T\dfrac{3R}{Na^2}$

Again, the force on the chain is simply the mass of the weighttimes gravitational acceleration, or $f = mg$

Thus, $mg = -k_B T\dfrac{3R}{Na^2}$

This is final expression]

4. Explain briefly the criteria for polymer solubility with example.

5. Explain briefly thethermodynamics of polymer solutions.

6. Explain Flory-Huggins theory.

7. Calculate entropy of polymer mixing.

8. Calculate enthalpy of polymer mixing

Answers

A

1. b	2. c	3. a	4. a	5. d	6. a	7. d	8. c	9. a	10. a

9

Preparation, Structure, Properties and Application of Polymers

Nowadays polymers are everywhere. It is deeply related in our day to day life. Researchers are continually working on polymer synthesis, polymer processing and to find the new properties of polymers. This type of study is very important both scientifically and industrially. It has a great impact on human life-style. Every polymer has importance in their respective application. Here their usual synthesis, well known structure, reported properties and wide range of application are discussed in a compact way.

Preparation of polyolefins, polystyrene ← styrene copolymers, polyvinyl chloride and related polymers, polyvinylacetate and related polymers, acrylic polymers, fluoropolymers, polyamides, phenol formaldehyde resins, polyurethanes, silicone polymers, polydienes polycarbonate and conducting polymers.

Hints : ⇒ First find out the monomer or monomers required for the polymer synthesis. ⇒ Then consider the polymerization processes (addition polymerization, condensation polymerization, coordination polymerization and co-polymerization) in which the monomer or monomers fit for polymerization. (Take help from chapter-3) ⇒ Then find out the reaction (mechanism). (Take help from chapter-3)

9.1 Polyolefins or Polyalkenes

▶ **Structures of polyolefins or polyalkenes :**

General structure

$$\left(\begin{array}{cc} H & H \\ | & | \\ C\!-\!C \\ | & | \\ H & R \end{array}\right)_n$$

Polyethylene
(R=H)

$$\left(\begin{array}{cc} H & H \\ | & | \\ C\!-\!C \\ | & | \\ H & H \end{array}\right)_n$$

Polypropylene
(R=CH$_3$)

$$\left(\!\!\begin{array}{cc} \text{H} & \text{H} \\ | & | \\ \text{C} - \text{C} \\ | & | \\ \text{H} & \text{CH}_3 \end{array}\!\!\right)_{\!n}$$

There are three types of polypropylene namely atactic polypropylene (aPP), isotactic polypropylene (iPP), syndiotactic polypropylene (sPP).

Polybutylene
(R=C$_2$H$_5$)

$$\left(\!\!\begin{array}{cc} \text{H} & \text{H} \\ | & | \\ \text{C} - \text{C} \\ | & | \\ \text{H} & \text{C}_2\text{H}_5 \end{array}\!\!\right)_{\!n}$$

Polypropylethylene
(R=C$_3$H$_7$)

$$\left(\!\!\begin{array}{cc} \text{H} & \text{H} \\ | & | \\ \text{C} - \text{C} \\ | & | \\ \text{H} & \text{C}_3\text{H}_7 \end{array}\!\!\right)_{\!n}$$

Poly(isopropylethylene)
R=CH(CH$_3$)(CH$_3$)

$$\left(\!\!\begin{array}{cc} \text{H} & \text{H} \\ | & | \\ \text{C} - \text{C} \\ | & | \\ \text{H} & \text{CH(CH}_3)(\text{CH}_3) \end{array}\!\!\right)_{\!n}$$

Poly(isobutylethylene)
R=CH$_2$CH(CH$_3$)(CH$_3$)

$$\left(\!\!\begin{array}{cc} \text{H} & \text{H} \\ | & | \\ \text{C} - \text{C} \\ | & | \\ \text{H} & \text{CH}_2\text{CH(CH}_3)(\text{CH}_3) \end{array}\!\!\right)_{\!n}$$

▶ **Preparation of polyolefins or polyalkenes :** Generally by polyolefins we understand polyethylene, polypropylene, polybutene, polypentene and their copolymers but widely used are polyethylene and polypropylene. Polyolefins are synthesized by addition or radical polymerization using specific radical forming initiator and/or Ziegler-Natta or Metallocene as catalysts. High Density Polyethylene (HDPE) has been prepared by polymerization of ethylene at 30-80°C temperature, 2-6 atmosphere pressure and using Ziegler-Natta or metallocene as catalyst. Low Density Polyethylene (LDPE) has been prepared by polymerization of ethylene at 80-350°C temperature, 1000-3000 atmosphere pressure and using traces of oxygen or peroxide as initiator. Polypropylene also has been prepared by polymerization of propylene at 20-50°C temperature, 30-60 atmosphere pressure and using Ziegler-Natta or metallocene as catalyst.

▶ **Properties of polyolefins or polyalkenes :** Depending on synthesis procedure the prepared polyethylene have a range of density. Again according to the density, polyethylene mainly grades into High-density polyethylene (HDPE) and Low-density polyethylene (LDPE). Commercially polyethylenes are available in the name of Ultra-high-molecular-weight polyethylene (HMWPE),

High-molecular-weight polyethylene (HMWPE), High-density polyethylene (HDPE), High-density cross-linked (cross-linkable) polyethylene (HDXLPE), Medium-density polyethylene (MDPE), Linear Medium-density polyethylene (LMDPE), Cross-linked (cross-linkable) polyethylene (PEX or XLPE), Low-density polyethylene (LDPE), Linear low-density polyethylene (LLDPE), Very-low-density polyethylene (VLDPE) and Ultra-low-molecular-weight polyethylene (PE-WAX).

High density polyethylene (HDPE) has 0.2 - 0.4 N/mm^2 tensile strength, density 0.94-0.97 g/cc, solubility parameter 17-20 Mpa$^{1/2}$, glass transition temperature (T_g) –30°C, refractive index 1.48, melting point ~ 400K, low temperature resistance, outstanding electrical insulating properties and low water absorption. It is highly crystalline (> 90% crystalline) and less amorphous.

Low density polyethylene (LDPE) has tensile strength: 0.2-0.4 N/mm^2, density 0.91-0.94 g/cc, solubility parameter 14-17 Mpa$^{1/2}$, glass transition temperature (T_g) –80°C, refractive index 1.47, melting point ~350K, low temperature resistance, outstanding electrical insulating properties and low water absorption. It is low crystalline (<50% crystalline) and highly amorphous.

Polypropylene has tensile strength 0.13-0.18 N/mm^2, density 0.85 - 0.88 g/cc, solubility parameter 15-18 Mpa$^{1/2}$, glass transition temperature (T_g) –20°C, refractive index 1.49, melting point ~ 430K, highly flammable, sensitive to microbial attacks, low temperature resistance, resistance to environmental stress cracking.

The properties will vary with synthesis procedures, reactant concentrations, temperature and all other related parameters. Thus it will be confusing and sometimes give wrong interpretation when we mention the specific properties. It is not correct to write PVA has T_g at 60°C. So when we want to write T_g of PVA, we have to mention synthesis procedure and in which instrument T_g was measured. For this type of controversy specific properties are omitted in the properties discussion.

▶ **Application of polyolefins or polyalkenes :** Due to its high density and crystalline structure high density polyethylene (HDPE) is applicable in food packaging, in making of trays, milk bottles, garbage containers, pipes and fittings (gas, water, sewage and drainage), cable protection, steel pipe coating, fuel tanks, wiring and cables. Whereas low density polyethylene (LDPE) is applicable in pharmaceutical bottles, caps and closures, frozen food packaging, laminations, water pipes and hoses pipes. And polypropylene is used in automotive industry, furniture industry, Laboratory apparatus, medical devices and textiles. But keep in mind that specific polyolefins have specific application.

9.2 Polystyrene and Styrene Copolymers

▶ **Structures of polystyrene and styrene copolymers:**

General structure

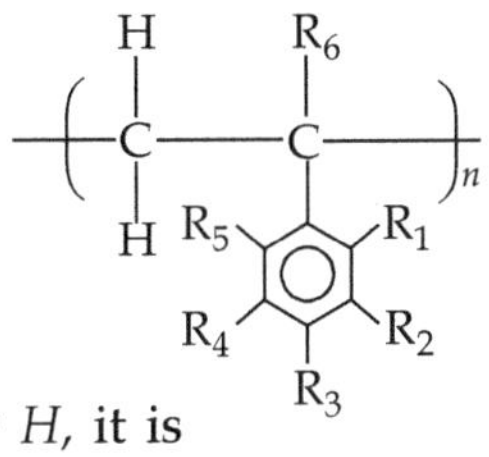

When $R_1 = R_2 = R_3 = R_4 = R_5 = H$, it is

Polystyrene

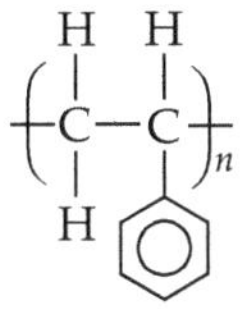

There are three types of polystyrene namely atactic polystyrene (aPP), isotactic polystyrene (iPP), syndiotactic polystyrene (sPP).

Poly(4-methylstyrene)

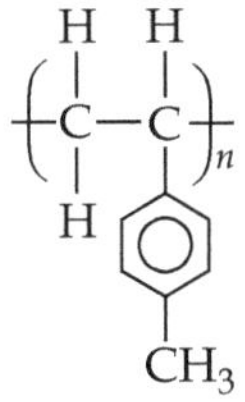

Poly(α-methylstyrene)

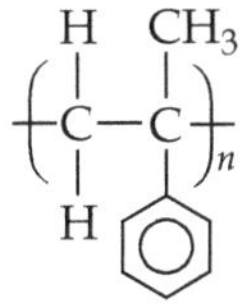

Poly(2-methylstyrene)

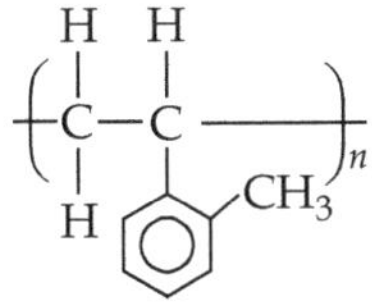

Poly(2-methoxystyrene)

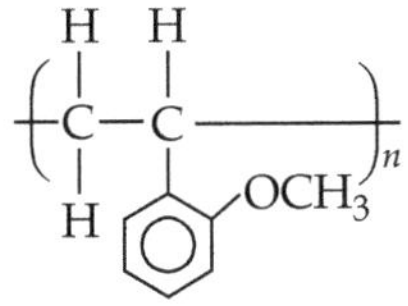

Poly(4-tert-butylstyrene)

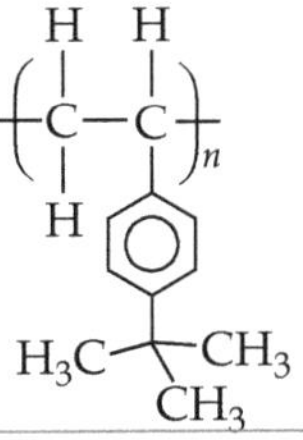

▶ **Styrene Copolymers :**

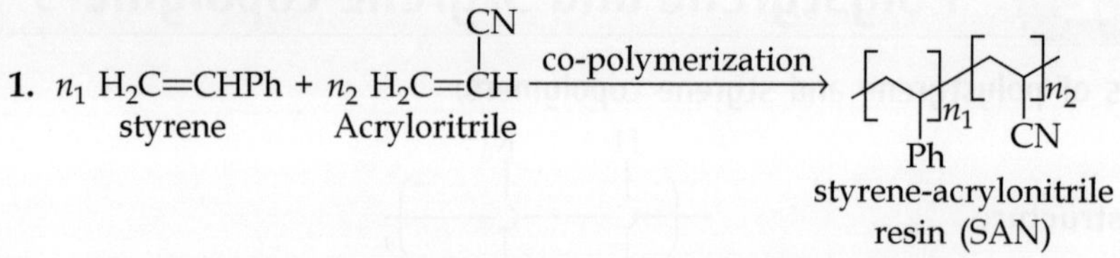

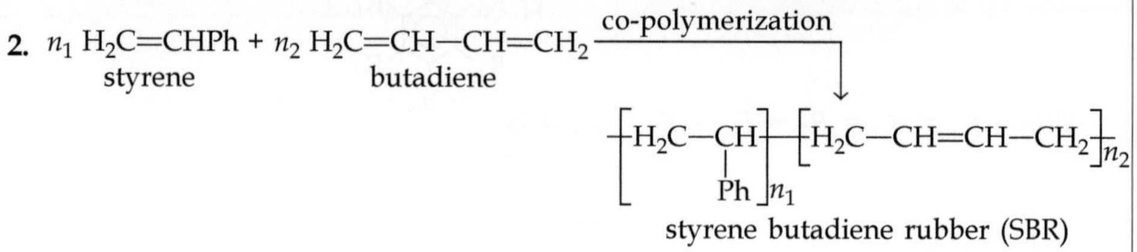

1. n_1 H$_2$C=CHPh + n_2 H$_2$C=CH(CN) $\xrightarrow{\text{co-polymerization}}$ styrene-acrylonitrile resin (SAN)
   styrene   Acryloritrile

2. n_1 H$_2$C=CHPh + n_2 H$_2$C=CH—CH=CH$_2$ $\xrightarrow{\text{co-polymerization}}$
   styrene   butadiene

   styrene butadiene rubber (SBR)

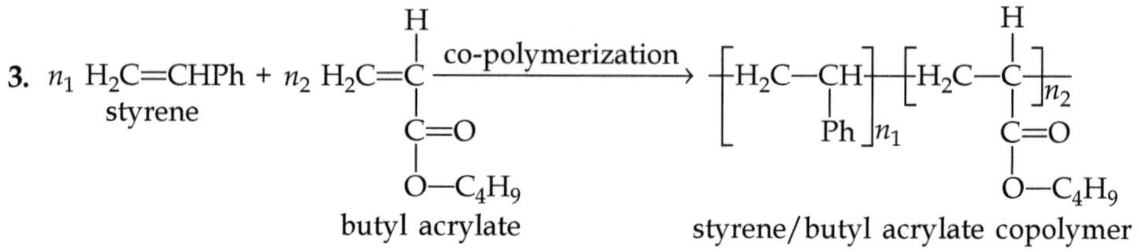

3. n_1 H$_2$C=CHPh + n_2 H$_2$C=C (with H, C=O, O—C$_4$H$_9$) $\xrightarrow{\text{co-polymerization}}$
   styrene   butyl acrylate

   styrene/butyl acrylate copolymer

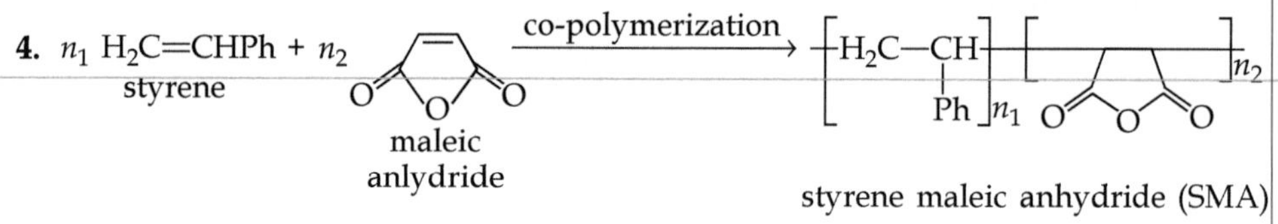

4. n_1 H$_2$C=CHPh + n_2 maleic anlydride $\xrightarrow{\text{co-polymerization}}$
   styrene

   styrene maleic anhydride (SMA)

* Prepared by living polymerization.

▶ **Preparation of polystyrene:** Polystyrene was synthesized by polymerization of styrene using benzoyl peroxide or Ziegler-Natta as the initiator. The mixture of styrene and benzoyl peroxide were heated at 90°C to get polystyrene.

▶ **Properties of polystyrene:** It has tensile strength : 34 MPa, density ~1.05 g/cc, solubility parameter 17-21 Mpa$^{1/2}$, glass transition temperature (T_g) ~ 100°C, melting point ~ 270°C, refractive index 1.59, rigid, brittle, gamma radiation resistance and UV resistance.

▶ **Application of polystyrene:** It is applicable in packaging industry, thermocol slabs, Disposable razor, medical devices and CD case.

9.3 Poly(vinyl chloride) and Related Polymers

▶ **Structures of poly(vinyl chloride) and related polymers:**

General structure

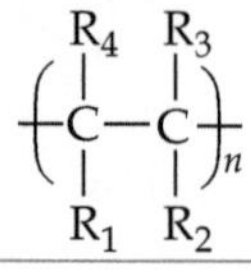

$$\left(\begin{array}{c} R_4 \ R_3 \\ | \ \ | \\ C—C \\ | \ \ | \\ R_1 \ R_2 \end{array} \right)_n$$

[R$_1$, R$_2$, R$_3$, R$_4$ = H, F, Cl, Br but one component must be Cl]

Poly(vinyl chloride)

$$\left(\!\!\begin{array}{c} \text{H} \quad \text{H} \\ | \quad\ | \\ \text{C}\!-\!\text{C} \\ | \quad\ | \\ \text{H} \quad \text{Cl} \end{array}\!\!\right)_n$$

Poly(vinylidene chloride)

$$\left(\!\!\begin{array}{c} \text{H} \quad \text{Cl} \\ | \quad\ | \\ \text{C}\!-\!\text{C} \\ | \quad\ | \\ \text{H} \quad \text{Cl} \end{array}\!\!\right)_n$$

Poly(chlorotrifluoroethylene)

$$\left(\!\!\begin{array}{c} \text{F} \quad \text{F} \\ | \quad\ | \\ \text{C}\!-\!\text{C} \\ | \quad\ | \\ \text{F} \quad \text{Cl} \end{array}\!\!\right)_n$$

▶ **Preparation of poly(vinyl chloride)** : Poly(vinyl chloride) was prepared by polymerization of vinyl chloride monomer using mainly suspension polymerization method. Emulsion polymerization and bulk polymerization are also used but rare or in few.

▶ **Properties of poly(vinyl chloride)** : It has elastic modulus ~ 2.4 GPa, density ~1.38 g/cc, solubility parameter 18-22 Mpa$^{1/2}$, glass transition temperature (T_g) ~ 70°C, melting point ~ 250°C, refractive index 1.53, rigid and flexible.

▶ **Application of poly(vinyl chloride)** : It is used in construction (sliding, window frames, and door frames), pipes and fittings (gas, water, sewage and drainage), clothing (coats, shoes and jackets), healthcare and flooring etc.

9.4 Poly(vinyl acetate) and Related Polymers

▶ **Structures of poly(vinyl acetate) and related polymers :**

General structure

$$\left(\!\!\begin{array}{c} \text{H} \quad \text{H} \\ | \quad\ | \\ \text{C}\!-\!\text{C} \\ | \quad\ | \\ \text{H} \quad \text{OCOR} \end{array}\!\!\right)_n$$

Poly(vinyl acetate)

$$\left(\!\!\begin{array}{c} \text{H} \quad \text{H} \\ | \quad\ | \\ \text{C}\!-\!\text{C} \\ | \quad\ | \\ \text{H} \quad \text{OCOCH}_3 \end{array}\!\!\right)_n$$

Poly(vinyl benzoate)

$$\left(\!\!\begin{array}{c} \text{H} \quad \text{H} \\ | \quad\ | \\ \text{C}\!-\!\text{C} \\ | \quad\ | \\ \text{H} \quad \text{OCOPh} \end{array}\!\!\right)_n$$

Poly(vinyl formate)	$\left(\!\!\begin{array}{cc} H & H \\ -C & -C- \\ H & OCHO \end{array}\!\!\right)_{\!n}$
Poly(vinyl butyrate)	$\left(\!\!\begin{array}{cc} H & H \\ -C & -C- \\ H & OCOCH_2CH_2CH_3 \end{array}\!\!\right)_{\!n}$
Poly(vinyl stearate)	$\left(\!\!\begin{array}{cc} H & H \\ -C & -C- \\ H & OCOC_{17}H_{35} \end{array}\!\!\right)_{\!n}$

▶ **Preparation of poly(vinyl acetate) and related polymers:** These types of polymers are prepared by conventional free-radical polymerization method.

▶ **Properties of poly(vinyl acetate) and related polymers:** poly(vinyl acetate) is a completely atactic, highly branched, amorphous thermoplastic and T_g is 30-45°C depending on the molecular weight. Poly(vinyl acetate) and related polymers are UV and oxidation resistance, good adhesive.

▶ **Application of poly(vinyl acetate) and related polymers :** Poly(vinyl acetate) and related polymers are used in the adhesive market. These are widely used as adhesive for coating industry, adhesives for porous materials, particularly for wood, paper, and cloth. It is mainly used as wood glue, carpenter's glue and school glue.

9.5   Acrylic Polymers

▶ **Structures of acrylic polymers :**

General structure	$\left(\!\!\begin{array}{cc} H & H \\ -C & -C- \\ H & COOR \end{array}\!\!\right)_{\!n}$
Poly(acrylic acid)	$\left(\!\!\begin{array}{cc} H & H \\ -C & -C- \\ H & COOH \end{array}\!\!\right)_{\!n}$
Poly(methyl acrylate)	$\left(\!\!\begin{array}{cc} H & H \\ -C & -C- \\ H & COOCH_3 \end{array}\!\!\right)_{\!n}$
Poly(ethyl acrylate)	$\left(\!\!\begin{array}{cc} H & H \\ -C & -C- \\ H & COOC_2H_5 \end{array}\!\!\right)_{\!n}$

Poly(benzyl acrylate)

$$\left(\!\!\begin{array}{cc} \overset{\text{H}}{\underset{\text{H}}{\text{C}}} & \overset{\text{H}}{\underset{\text{COOCH}_2\text{Ph}}{\text{C}}} \end{array}\!\!\right)_n$$

Poly(tert-butyl acrylate)

$$\left(\!\!\begin{array}{cc} \overset{\text{H}}{\underset{\text{H}}{\text{C}}} & \overset{\text{H}}{\underset{\text{COOC(CH}_3)(\text{CH}_3)(\text{CH}_3)}{\text{C}}} \end{array}\!\!\right)_n$$

▶ **Preparation of acrylic polymers :** These types of polymers are prepared by free-radical solution polymerization method.

▶ **Properties of acrylic polymers :** These polymers are transparent, resistance to breakage, and elastic. They are strong, tough, and lightweight materials.

▶ **Application of acrylic polymers :** PMMA is used as UV wavelength cutter (<300 nm). PMMA is also used in aquariums, aircraft windows, lighthouse lenses, eyeglass, hard contact lenses and in many medical devices. It was used for the roofing of the Munich Olympic Park.

9.6 Fluoropolymers

▶ **Structures of fluoropolymers:**

General structure

$$\left(\!\!\begin{array}{cc} \overset{\text{R}_4}{\underset{\text{R}_1}{\text{C}}} & \overset{\text{R}_3}{\underset{\text{R}_2}{\text{C}}} \end{array}\!\!\right)_n$$

[R_1 , R_2 , R_3 , R_4 = H, F, Cl, Br but one component must be F]

Poly(vinyl fluoride)

$$\left(\!\!\begin{array}{cc} \overset{\text{H}}{\underset{\text{H}}{\text{C}}} & \overset{\text{H}}{\underset{\text{F}}{\text{C}}} \end{array}\!\!\right)_n$$

Poly(vinylidene fluoride)

$$\left(\!\!\begin{array}{cc} \overset{\text{H}}{\underset{\text{H}}{\text{C}}} & \overset{\text{F}}{\underset{\text{F}}{\text{C}}} \end{array}\!\!\right)_n$$

Poly(tetrafluoroethylene)

$$\left(\!\!\begin{array}{cc} \overset{\text{F}}{\underset{\text{F}}{\text{C}}} & \overset{\text{F}}{\underset{\text{F}}{\text{C}}} \end{array}\!\!\right)_n$$

Poly(chlorotrifluoroethylene)

$$\left(\!\!\begin{array}{cc} \overset{\text{F}}{\underset{\text{F}}{\text{C}}} & \overset{\text{F}}{\underset{\text{Cl}}{\text{C}}} \end{array}\!\!\right)_n$$

▶ **Preparation of fluoropolymers:** These types of polymers are prepared by conventional free-radical polymerization method. But anionic ring opening polymerisation or UV-catalysed radical oxopolymerisation processes are also done.

▶ **Properties of fluoropolymers:** These polymers are chemical resistance, thermally stable and have low friction coefficients and low dielectric constants. These are expensive, have outstanding moisture barrier, interesting electronic properties and others.

▶ **Application of fluoropolymers:** These types of polymers are used in chemical processing, resistant components and coatings, pipes, fittings, linings, tapes, seals, filters, wire and cable insulation, laminates, waterproof and stain repellent clothing, architectural and carpet coatings, printing, cookware, fabrics, biomedical devices and many other fields.

9.7 Polyamides

▶ **Structures of polyamides :**

General structure $\left[\!-NH-R-\overset{\displaystyle O}{\underset{\displaystyle \|}{C}}-\right]_n$ or, $\left[\!-NH-R_1-NH-\overset{\displaystyle O}{\underset{\displaystyle \|}{C}}-R_2-\overset{\displaystyle O}{\underset{\displaystyle \|}{C}}-\right]_n$

Nylon 6 or Poly(caprolactam) $\left[\!-NH-(CH_2)_5-\overset{\displaystyle O}{\underset{\displaystyle \|}{C}}-\right]_n$

Nylon 8 or Polycapryllactam $\left[\!-NH-(CH_2)_7-\overset{\displaystyle O}{\underset{\displaystyle \|}{C}}-\right]_n$

Nylon 6, 6 or Poly(hexamethylene adipamide)

$$\left[\!-NH-(CH_2)_6-NH-\overset{\displaystyle O}{\underset{\displaystyle \|}{C}}-(CH_2)_4-\overset{\displaystyle O}{\underset{\displaystyle \|}{C}}-\right]_n$$

Nylon 10,10 or Poly(decamethylene sebacamide)

$$\left[\!-NH-(CH_2)_{10}-NH-\overset{\displaystyle O}{\underset{\displaystyle \|}{C}}-(CH_2)_8-\overset{\displaystyle O}{\underset{\displaystyle \|}{C}}-\right]_n$$

Kevlar or Polyaramide $\left[\!-NH-\!\!\bigcirc\!\!-NH-\overset{\displaystyle O}{\underset{\displaystyle \|}{C}}-\!\!\bigcirc\!\!-\overset{\displaystyle O}{\underset{\displaystyle \|}{C}}-\right]_n$

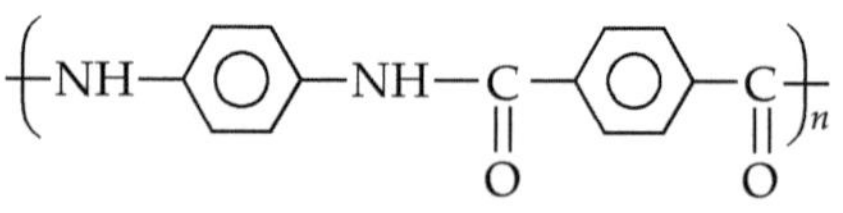

Nomex or Poly(*m*-phenylene terephthalamide)

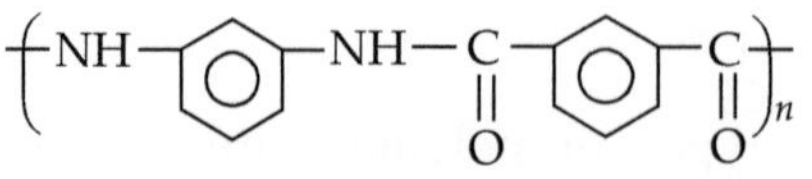

$$\left[\!-NH-\!\!\bigcirc\!\!-NH-\overset{\displaystyle O}{\underset{\displaystyle \|}{C}}-\!\!\bigcirc\!\!-\overset{\displaystyle O}{\underset{\displaystyle \|}{C}}-\right]_n$$

▶ **Preparation of polyamides :** These types of polymers are prepared by condensation polymerization method by the reaction of a diacid with a diamine or by ring-opening polymerization of lactams. An amide linkage is necessary for polymer formation.

▶ **Properties of polyamides :** These polymers are high performance aliphatic, semi-aromatic or fully aromatic thermoplastics. The aromatic polyamides have higher strength, better solvent, flame and heat resistance than the aliphatic amides. The aromatic polyamides, called aramids, are much more expensive and more difficult to produce. The two most important aromatic amides Kevlar and Nomex have excellent properties for many applications.

▶ **Application of polyamides :** These types of polymers are widely used in engineering and industrial applications. Nylons are used in insulating wire and cable, cooling fans, air circulator, valve and engine covers, wipers and speedometers, switches, sockets, plugs and antenna-mounting devices. Kevlar is used in cryogenics, combat helmets, ballistic face masks, and ballistic vests, motorcycle safety clothing, shoes, cycle tyre and many other areas. Nomex is used in firefighting equipment, Racing car driver's suits, Military pilots and aircrew suits and many safety equipment.

9.8 Phenol Formaldehyde Resins

▶ **Structures of phenol formaldehyde resins :**

General structure

Novolac

Bakelite

▶ **Preparation of phenol formaldehyde resins :** These types of polymers are prepared by step growth and/or cross-linked method by the reaction with phenol and formaldehyde. As phenol is ortho and para directing group so it reacts with formaldehyde giving several structures where formaldehyde can attach in any 1 centre or 2 centres or 3 centres of phenol ring (2, 4 and 6 substituted).

Novolac is prepared by the reaction of phenol with formaldehyde. Bakelite is prepared by heating Novolac or with excess formaldehyde.

▶ **Properties of phenol formaldehyde resins :** These types of polymers have good mechanical, electrical insulation, durable, and heat and flame resistance property.

▶ **Application of phenol formaldehyde resins :** Phenol formaldehyde resins are widely used as coating, adhesive, foaming material, billiard balls and laboratory countertops. Novolacs is used in high temperature resin, carbon brakes, photoresists and curing agent for epoxy resins. Bakelite is used in molding compound, adhesive or binding agent, varnish and protective coating.

9.9 Polyurethanes

▶ **Structures of polyurethanes :**

General structure

$$\left(\!\!\begin{array}{c} \text{C} \\ \| \\ \text{O} \end{array}\!\!-\text{NH}-\text{R}_1-\text{NH}-\!\!\begin{array}{c}\text{C}\\\|\\\text{O}\end{array}\!\!-\text{O}-\text{R}_2-\text{O}\!\!\right)_{\!n}$$

Poly[(1,4-butanediol)-*alt*-(4,4′-diphenylmethane diisocyanate)]

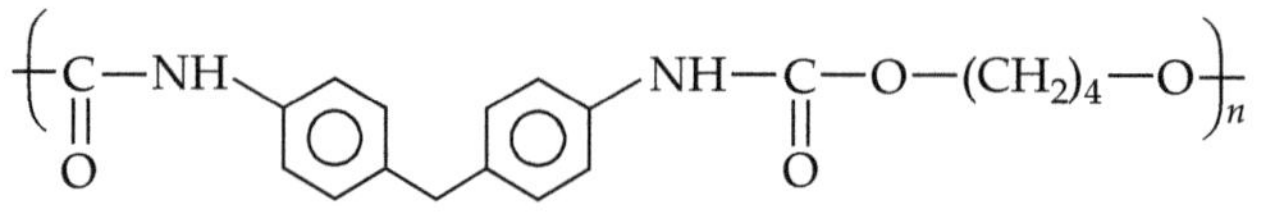

Poly[(ethylene glycol)-*alt*-(4,4′-diphenylmethane diisocyanate)]

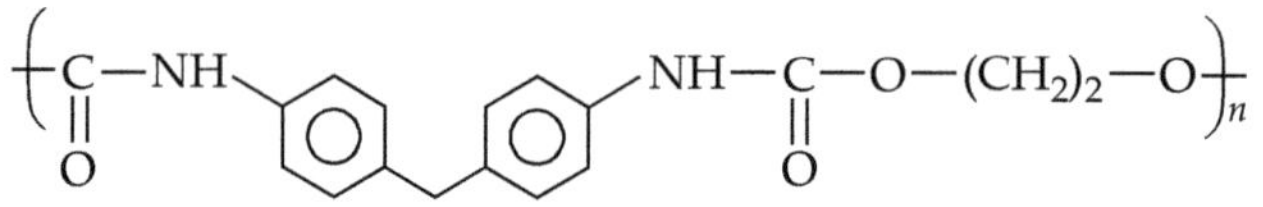

Poly[(diethylene glycol)-*alt*-(1,6-hexamethylene diisocyanate)]

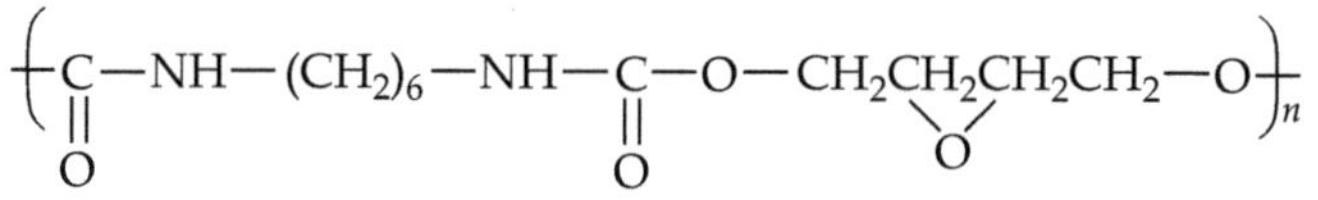

Poly[(tetraethylene glycol)-*alt*-(1,6-hexamethylene diisocyanate)]

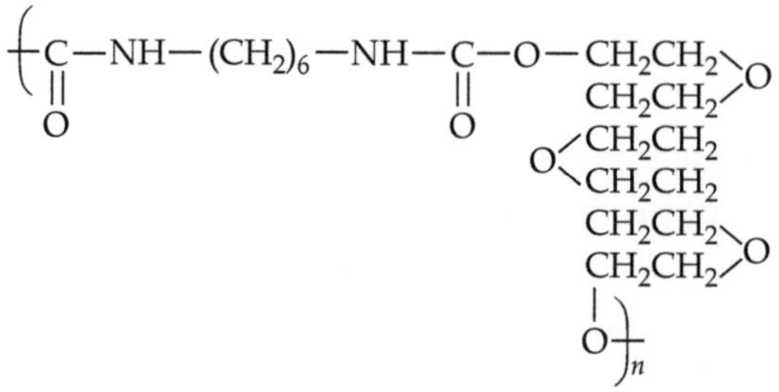

▶ **Preparation of polyurethanes :** These types of polymers are generally prepared by step-growth polyaddition reaction of diisocyanates and polyols in the presence of suitable catalysts and additives.

▶ **Properties of polyurethanes :** These types of polymers can be rigid or flexible, hard or soft and thermosetting or thermoplastic. The main chains have urethane linkages (-NH-C(O)-O-) in the backbone.

▶ **Application of polyurethanes :** Polyurethanes are the top classes of polymers with excellent properties which make them in wide range of applications. They are applicable in producing flexible and rigid foams like mattresses, carpet, cushions and armrests, insulations for commercial and residential buildings, tanks, pipes, water heaters, refrigerators, and freezers. Polyurethanes are also applicable as elastomeric products like bumpers, fenders, steering wheels, instrument and door panels, and gaskets for trunks, windows and windshields. They are also applicable in adhesives, coatings, sealants, and synthetic fibers.

9.10 Silicone Polymers

▶ **Structures of silicone polymers :**

General structure $\left(\!-O-\overset{\displaystyle R_1}{\underset{\displaystyle R_2}{Si}}\!-\right)_{\!n}$

Poly(dimethylsiloxane) (PDMS) $\left(\!-O-\overset{\displaystyle CH_3}{\underset{\displaystyle CH_3}{Si}}\!-\right)_{\!n}$

Poly(diethylsiloxane) (PDES) $\left(\!-O-\overset{\displaystyle C_2H_5}{\underset{\displaystyle C_2H_5}{Si}}\!-\right)_{\!n}$

Poly(methylphenylsiloxane) (PMPS) $\left(\!-O-\overset{\displaystyle Ph}{\underset{\displaystyle CH_3}{Si}}\!-\right)_{\!n}$

▶ **Preparation of silicone polymers :** Generally silicone polymers are prepared by hydrolysis of chlorosilanes. The resultant polymer is linear but it may be cyclic.

▶ **Properties of silicone polymers :** These types of polymers have low toxicity, low thermal conductivity and thermal stability and high resistance to weathering and many chemicals. It has good electrical insulation properties and high gas permeability.

▶ **Application of silicone polymers :** These types of polymers are widely used in engineering and industrial applications. They are good sealant, adhesive, surfactant and antifoam agent, paper coating material, lubricants, medicine, cooking utensils, and thermal and electrical insulator. Silicone oil, silicone grease, silicone rubber and silicone resins are few common forms of silicone polymers.

9.11 Polydienes

▶ **Structures of polydienes :**

General structure — and $-\left(R_1HC\!=\!CHR_2 \right)_n$

$$\left(\begin{array}{cc} R & H \\ | & | \\ C - C \\ | & | \\ H & CH\!=\!CH_2 \end{array} \right)_n$$

Poly(1,2-butadiene)

$$\left(\begin{array}{cc} H & H \\ | & | \\ C - C \\ | & | \\ H & CH\!=\!CH_2 \end{array} \right)_n$$

Poly(1,4-butadiene) $-\left(CH_2HC\!=\!CHCH_2 \right)_n$

Poly(1-ethyl-1,4-butadiene) $-\left(CH(C_2H_5)HC\!=\!CHCH_2 \right)_n$

Polycyclopentene $-\left(CH_2CH_2CH_2HC\!=\!CH \right)_n$

Polyisoprene $-\left(CH_2(CH_3)C\!=\!CHCH_2 \right)_n$

▶ **Preparation of polydienes :** These types of polymers are generally prepared by free-radical, ionic and ion-coordination reaction in the presence of suitable catalysts (mainly Ziegler-Natta catalysts).

▶ **Properties of polydienes :** These types of polymers have very excellent elastic properties, low glass transition temperature and good UV stability. It has chemical resistance (mainly oxygen and ozone), low temperatures resistance. Sometimes polydienes contain unsaturated sites which can further vulcanized to produce an elastomeric network. Styrene-butadiene rubber, polybutadiene, nitrile-butadiene rubber, polychloroprene and polyisoprene are the common diene elastomers which have excellent properties for many applications.

▶ **Application of polydienes :** These types of polymers are widely used in industrial applications. It is used in automobile tires, carpet backing, plastic, gloves, wetsuits, rubber hoses, golf balls, railway pads, bridge blocks, gaskets and defence equipments.

9.12 Polycarbonates

▶ **Structures of polycarbonates :**

General structure $\left(\!R\!-\!O\!-\!\underset{\underset{O}{\|}}{C}\!-\!O\!\right)_{n}$

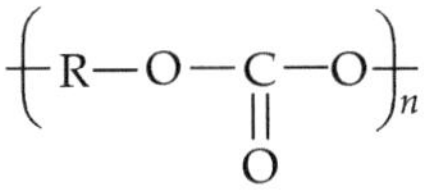

Poly(ethylene carbonate) $\left(\!CH_2CH_2\!-\!O\!-\!\underset{\underset{O}{\|}}{C}\!-\!O\!\right)_{n}$

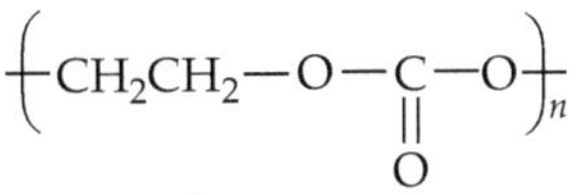

Poly(propylene carbonate) $\left(\!CH(CH_3)CH_2\!-\!O\!-\!\underset{\underset{O}{\|}}{C}\!-\!O\!\right)_{n}$

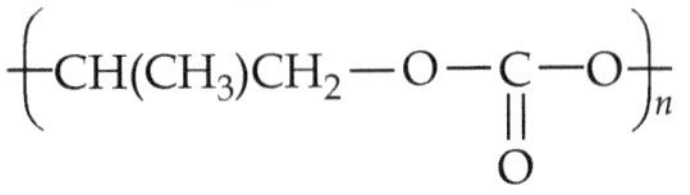

Poly(bisphenol A carbonate)

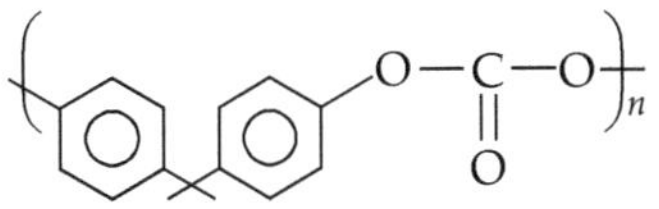

Poly(4,4′-thiodiphenylene carbonate)

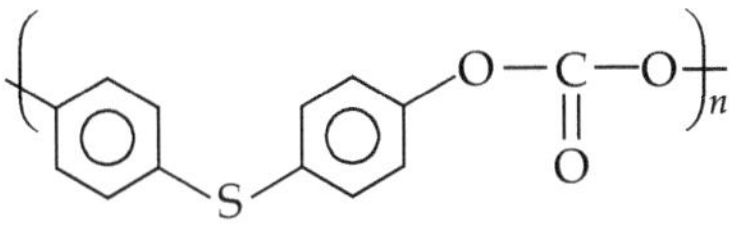

▶ **Preparation of polycarbonates :** These types of polymers are generally prepared by condensation polymerization between aromatic dihydroxy (mainly bisphenol A) or epoxy or similar compound and either carbonyl chloride or diphenyl carbonate.

▶ **Properties of polycarbonates :** The main chains have carbonate groups ("O"(C=O)"O") in the backbone. Polycarbonate is durable, impact-resistant and scratch-resistant material. It can be deformed without cracking or breaking, has a glass transition temperature above room temperature.

▶ **Application of polycarbonates :** These types of polymers are widely used in construction materials (safety helmets, power tools, appliance housings, glazing for windows, doors and face shields), medical equipment, mobile phone, compact discs, sun glasses, automotive, aircraft, and security components.

9.13 Conducting Polymers

▶ **Structures of conducting polymers:**

Polyacetylene $\left(\!\underset{\underset{H}{|}}{\overset{\overset{H}{|}}{C}}\!=\!C\!\right)_{n}$

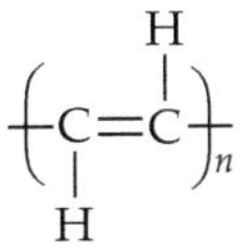

Polyaniline	$\left(\!\!\!\begin{array}{c}\\ \text{benzene ring} \end{array}\!\!-\!\overset{\overset{\displaystyle H}{\mid}}{N}\!\right)_{\!n}$
poly(*p*-phenylenesulphide)	$\left(\!-\!\text{benzene}\!-\!S\!-\!\right)_{\!n}$
Polypyrrole	$\left(\!\!\begin{array}{c} \text{pyrrole ring} \\ \underset{\overset{\mid}{H}}{N} \end{array}\!\!\right)_{\!n}$
Polythiophene	$\left(\!\!\begin{array}{c} \text{thiophene ring} \\ S \end{array}\!\!\right)_{\!n}$
Polyphenylene	$\left(\!-\!\text{benzene}\!-\!\right)_{\!n}$
Poly(*p*-phenylene vinylene)	$\left(\!-\!\text{benzene}\!-\!CH\!=\!CH\!-\!\right)_{\!n}$

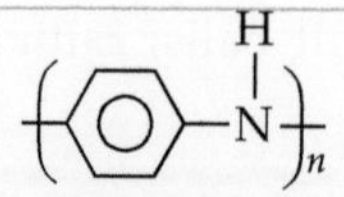

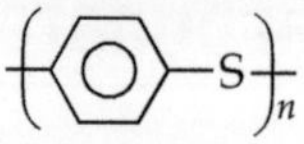

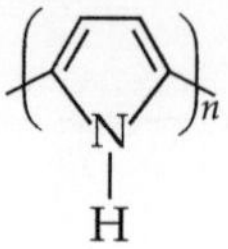

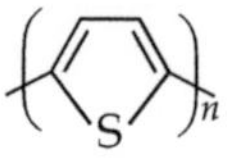

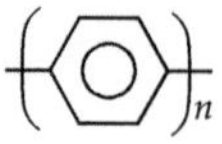

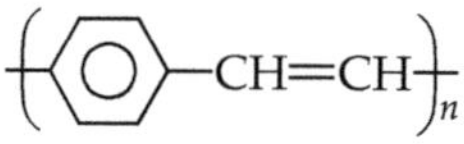

▶ **Preparation of conducting polymers :** Depending on monomers' nature conducting polymers have been prepared in various way. It can be prepared in aqueous medium or organic medium, in presence of air or in presence of inert gas, solution or bulk or interfacial or solid state polymerization. But mainly radical polymerization method is used. As for example polyaniline and polypyrrole can be prepared in aqueous acidic medium but polythiophene have to be prepared in chloroform or organic medium.

▶ **Properties of conducting polymers :** It is named as conducting polymers because these types of polymers can conduct charge or electricity. But it can be conducting, semi-conducting or insulating in nature. It is cheap, easy to synthesize, has nice optical properties but mechanical properties are not good. Fillers are added to improve the mechanical properties.

▶ **Application of conducting polymers :** These types of polymers are used in light emitting diodes, organic solar cells, electronic circuits, corrosion resistance, electromagnetic shielding, molecular electronics, supercapacitors, electrochromic devices, actuators, and chemical and biosensors.

When any learner (student) wants to enter into polymer field, overall knowledge of all polymers gives the idea how polymers are doing in the world. The previous part is written in such a way but when anyone (researcher) wants to know about one specific polymer till to date, comprehensive study (though complete study is not feasible because day by day it is upgrading) is required. Let have a look if anyone want to know about conducting polymer polyaniline. Keep in mind that it is only one specific polymer of conducting polymer family and the complete polymer family is vast.

- **History of polyaniline :** Polyaniline (PANI) is a conducting electroactive polymer having complex dynamic structures for intelligent materials research. Polyaniline have been known as "aniline blacks" for more than a century and used as cotton dyes since 1860s. Systematic studies of their properties prior to World War 1 lead Green and Woodhead to propose that aniline have octamer structure with several oxidation states. Only random studies of polyaniline were undertaken over the following 70 years. MacDiarmid *et al.* discovered electrical conductivity for polyaniline emeraldine salt form in 1980. Polyaniline is a novel-conducting polymer due to its straightforward polymerization, chemical stability, acid doping/base dedoping process and good environmental stability. Depending on morphology and electronic state of polyaniline, it is used in chemical and biological sensor, asymmetric polymer membranes, electronic devices, nonvolatile memory devices, light weight battery electrodes, electromagnetic shielding devices, anticorrosion coating, chiral separation and surface modified electrodes and neuron devices. However, its physical, thermal and mechanical properties are not very good and find difficulties in many applications. To improve its mechanical, thermal and other physical properties different types of nanomaterials (fillers) are introduced into the polymer. It is interesting to both academic and industrial researchers from different disciplines of science and technology.

- **Synthesis of polyaniline :** Polyaniline has been synthesized by the following procedures (i) chemical polymerization (ii) Electrochemical polymerization (iii) Photochemically initiated polymerization (iv) Enzyme-catalyzed polymerization and (v) polymerization employing electron acceptors.

  i. **Chemical polymerization :** The emeraldine salt (ES) form of PANI were synthesized easily as amorphous or partly crystalline black or green precipitate by oxidative polymerization of aniline in presence of an oxidizing agent in the acidic solution. Different types of oxidizing agent has been used namely ammonium, persulfate, potassium permanganate, potassium iodide, ferric chloride, copper perchlorate, potassium dichromate etc. Ammonium persulfate (APS) in aqueous HCl media is widely used rather than other ones. The final form of PANI can be controlled by changing (i) nature of the medium (ii) mole ratio of oxidant to aniline (iii) duration of the reaction time (iv) reaction temperature in the polymerization process. MacDiarmid *et al.* synthesized PANI and characterized it with respect to its chemical structure and properties e.g. conductivity, doping level and molecular weight etc. In the synthesis of PANI the acidity of the polymerization medium has strong influence on the electrical conductivity. At pH < 4 the electrical conductivity and reaction yield remained unaffected but the viscosity average molecular weight of the product (measured at 25°C using 0.1% w/w solution in concentrated H_2SO_4 in ubbelohde viscometer) is dependent on the acidity. Various types of acids were used for the preparation of PANI but the most effective is HCl. The use of H_2SO_4 is avoided since a thin film of acid is left even after drying the polyaniline powder in vacuum. Cao *et al.* showed that the yield and electrical conductivity were not sensitive to the acid used. But the viscosity is strongly dependent on the nature of the acid used, because of difference in molecular weight of counter ions. However, Cao et al. found that the molecular weight of PANI was dependent upon the pH of the medium. For the values of 'r' (initial mole ratio of oxidant / aniline) less than 1.15, the conductivity, yield, elemental composition and degree of oxidation of the resulting PANI are essentially independent of r. But for $r > 1.5$ results in over oxidation which decreases both the yield and conductivity of PANI. Cao *et al.* proposed that the method of addition of oxidant is an important factor especially as far as the molecular weight of the polymer concerned. Not only the addition procedure of oxidant changes the molecular weight, such changes are also found when the temperature of the reaction media is changed. The optimum reaction temperature has been found to be in between 0 to 5°C and a slow addition of oxidant has also been recommended. In comparison of various oxidant used, APS and $K_2Cr_2O_7$ give high yield, high electrical conducting and high molecular weight PANI, KIO_3 and $FeCl_3$ yield very low molecular weight PANI and $KMnO_4$, $FeCl_3$, $KClO_3$ etc. gives very low yield as well as low molecular weight and electrical conducting PANI. The time of polymerization also plays an important role in PANI synthesis. Armes *et al.* prepared their samples in a range of 10 hrs to 90 hrs. But Cao *et al.* showed that longer polymerization time yields the low viscosity material *i.e.* low molecular weight product. They proposed 4 hrs is the optimum time at 0°C temperature keeping other variables in standard condition. The longer period does not give higher yield rather lead to chain scission, decrease the molecular weight due to slow hydrolysis. In conclusion the polymerization is a relatively rapid one and is completed within 4 hrs. It was found that the nature of aniline polymerization is an autocatalytic one. Tzou *et al.* proposed that polymerization of aniline occurs both in a homogeneous and heterogeneous medium in aqueous acid solution. The catalytic effect of the polyaniline formed by autocatalytic reaction occurs mainly in the heterogeneous phase. The high molecular weights of polyaniline in aqueous solutions might be explained by the fact that oligomers have precipitated (due to extremely low solubility in water) as active seeds and subsequent polymerization will occur preferentially on these seeds because of the autoacceleration reaction in the heterogeneous phase. But PANI oligomers remain in the solution phase for a long period of time when polymerizations done in organic solvents as the oligomers have higher solubility in organic solvents. In this case the auto-acceleration reaction may also occur in the homogeneous phase and a greater number of seeds will be formed in organic phase resulting in lower molecular weight. This is analogous to typical ionic polymerization reactions with higher concentration of initiator. Since more initiator is added, more active centres will be generated and fewer monomers will be able to grow onto one single active centre resulting lower molecular weight PANI. The emeraldine salt (ES) produced can be easily converted to emeraldine base (EB) by treatment with NH_4OH. This base material is soluble in NMP, DMSO etc. The base material can be redoped by stirring with acidic (*e.g.* HCl) solution for 2 hrs.

ii. **Electrochemical polymerization :** Electrochemical polymerization of aniline were done by i) constant current or galvanostatic mode ii) constant potential or potentiostatic mode and iii) by potential cycling methods. In the galvanostatic mode two electrode system are employed using a current density of 1 mA / sq. cm and they dipped in an electrolytic solution contain aniline (oxidation potential = 0.75V). In this process PANI is deposited as film form on Pt-electrode surface, where as in case of constant potential mode PANI is collected as powders. Under potentiostatic mode the potential applied is in the range of 0.7 to 1.2 V (versus SCE). In case of potential cyclic mode, the potential is in between – 0.2 V to 0.7 – 1.2 V (versus SCE). From scanning electron microscopy it is concluded that in this process much more homogenous product is obtained. The sweep rate varies between 10 and 100 mVs^{-1} gives good quality film. The potential cyclic modes give homogeneous oxidized cationic form of polyaniline film and the overall charge balance is achieved by incorporation of counter anion from the electrolytic solution. The counter anions used are F^-, Cl^-, ClO_4^-, SO_4^{-2}, BF_4^- etc. Platinum or conducting glasses are used as electrode materials for the anodic oxidation of aniline. Besides these electrodes some metals which are not oxidized readily than aniline such as Fe, Cu and Au have also been used. Sometime graphite, vitreous carbon and *n*-type silicon are used. Usually electrochemical polymerization is carried out in aqueous solution. But aprotic solvent may also be used e.g. propylene carbonate solution containing an organic acid (CF_3COOH) and lithium perchlorate as electrolyte. Wei et al. have shown that the mass of polyaniline deposited on a platinum electrode is controlled with the polymer anodic peak current that is recorded during the polymerization. The rate of polymer deposition is therefore monitored by the increase in the anodic peak current of various concentration of aniline. Several groups have observed with potentiostatic techniques that the initial formation (or nucleation) of PANI on the electrode surface is slow, but once the PANI is formed, it is accelerated the polymer growth greatly and the current increased proportionally with the second power of time. This is called 'auto-acceleration' or 'self–catalyzing' process in the electrochemical polymerization of aniline. Electro-polymerization also has certain advantages over chemical polymerization. They are (i) the resulting product is clean and does not required to be extracted from the monomer/oxidant/solvent mixture (ii) this method offers the possibility of coupling with physical spectroscopic techniques such as visible, I.R., ellipsometry and conductometry for in situ characterization.

iii. **Photochemically initiated polymerization:** Kobayashi *et al.* reported the photochemically initiated polymerization of aniline by irradiation with visible light of either bilayer or single layer films containing $[Ru(bipy)_3]^{2+}$ (as initiator) and methylviologen (MV^{2+}) as sacrificial oxidant. Irradiation of the $[Ru(bipy)_3]^{2+}$ near its absorption band at 452 nm leads to the generation of the triplet excited state *$[Ru(bipy)_3]^{2+}$ species. Electron transfer between the *$[Ru(bipy)_3]^{2+}$ and MV^{2+} generates the strong oxidant $[Ru(bipy)_3]^{3+}$ responsible for the oxidation/polymerization of aniline to polyaniline (PANI). It was, however, observed that small amounts of aniline dimer (e.g. 10^{-3} M) were necessary in the aniline monomer (0.3 M) substrate solution to achieve photopolymerization. This was attributed to the lower oxidation potential of the dimer compared with aniline, with the dimeric radical cation formed in the initial oxidation step then reacting with aniline monomer to give an even more readily oxidized trimer, eventually leading to PANI. Leon et al. have explored the use of both $[Ru(bipy)_3]^{2+}$ and $[Ru(biphen)_3]^{2+}$ as polymerization photoinitiators for the bulk synthesis of emeraldine salts. They found that by employing the cobalt (III) complexes $[CoCl(NH_3)_5]^{2+}$ and $[Co(H_2O)(NH_3)_5]^{3+}$ as the sacrificial oxidants rather than methylviologen. Aniline can be readily photo-polymerized to $PANI.HNO_3$ and PANI.HCSA (HCSA = camphor-10-sulfonic acid) salts without the need to add aniline dimer. Polymerization of substituted PANI could be similarly achieved, e.g. yielding poly(*n*-butylaniline).HNO_3 and poly(2,5-dimethoxyaniline).HNO_3 salts.

iv. **Enzyme-catalyzed polymerization :** The enzymes such as horseradish peroxidase (HRP) used as catalyst for the synthesis of PANI with oxidants such as peroxide. This method is environmental friendly (the H_2O_2 oxidant is converted into water). In the early studies PANI formed have low molecular weight and extensive chain branching. Samuelson et al. have overcome this problem for the HRP catalyzed synthesis of PANI through an elegant approach using polyelectrolytes such as polystyrenesulfonate (PSS) as templates in the reaction mixture to induce more structural order during polyaniline growth. The PSS plays three roles in this approach : (i) Serving as a template that aligns the aniline monomers prior to polymerization so as to promote the desired head-to-tail coupling, (ii) Providing the counter ions for doping the synthesized PANI to the conducting emeraldine salt form and (iii) Imparting water "solubility" to the PANI.PSS product. Another advantage of this enzyme-catalyzed route to colloidal PANI salts is the considerably higher pH that can be employed compared with the previous chemical and electrochemical polymerization methods. HRP immobilized on chitosan powder as a solid support has also been found to catalyze the H_2O_2 oxidation of aniline to a similar PANI.PSS product, opening up the prospect of enzyme reuse and the design of enzyme reactors for PANI synthesis. The mild (pH 4.3) conditions for these HRP-catalyzed synthesis have led to the similar employment of more-delicate biological polyelectrolytes such as DNA as the aligning templates. In this case, electrostatic interactions between the DNA phosphate groups and protonated aniline monomers prior to polymerization are believed to provide the preferential alignment leading to para-directed coupling of the aniline units. Again the binding of DNA to polyaniline in the PANI.DNA product leads to a remarkable reversible change in the conformation of the DNA chains.

v. **Polymerization employing electron acceptors:** Kuramoto *et al.* used an organic electron acceptor dichlorodicyanobenzoquinone (DDQ) for the polymerization of dodecylbenzenesulfonic acid (DBSA) salts of aniline or 2-methoxyaniline in chloroform solvent. The PANI.DBSA product was isolated by the addition of excess methanol and acetone to the polymerization mixture, simultaneously removing oligomers and other organic by products. However the weaker organic electron acceptors chloranil and TCNQ showed no oxidative polymerization activity. These researchers extended this approach to the synthesis of optically active polyanilines with the polymerization of aniline and its substituted analogues 2-methoxyaniline, 2-ethoxyaniline and *o*-toluidine with DDQ in mixed chloroform/THF solvent.

9.14 Properties of Polyaniline

Structure : Polyaniline is among the oldest known synthetic polymers. More than 100 years ago Letheby (1862) obtained a dark brown precipitate from the anodic oxidation of aniline on a pt-electrode in aqueous H_2SO_4. From the mid 1980 there was a considerable interest in this family of polymers. PANI exhibits number of oxidation states. The different oxidation states of PANI exhibit different colour and electrical properties with the conductivity ranging from 10^{-11} Scm^{-1} to 10^2 Scm^{-1}. Jozefowicz and coworkers observed in 1968 that the conductivity of PANI could be increased by orders of magnitude by decreasing the pH of the preparation medium. Since then PANI was subjected to intense characterization through its structure. Green and Woodhead proposed a linear octameric structure for the product obtained by chemical oxidation of aniline. The following figure shows the structure and some physicochemical properties of aniline octameric sequence as proposed by them. The description of the repeat unit of PANI as an octameric sequence is the best choice because it allows a straight forward distinction

▶ **Leucoemeraldine**

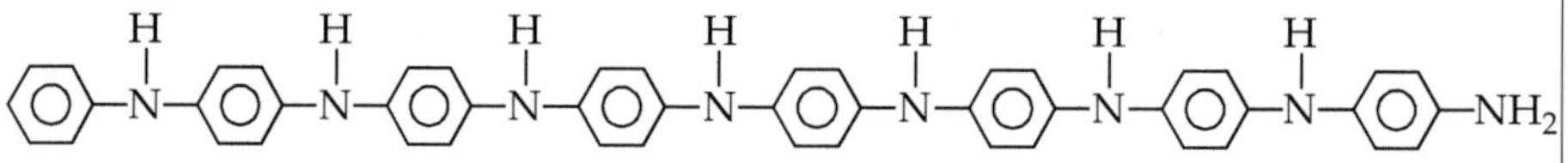

Pale-brown, amorphous, insoluble, high melting, slowly oxidized to protoemeraldine, more rapidly when warm in air.

▶ **Protoemeraldine (monoquinonoid state)**

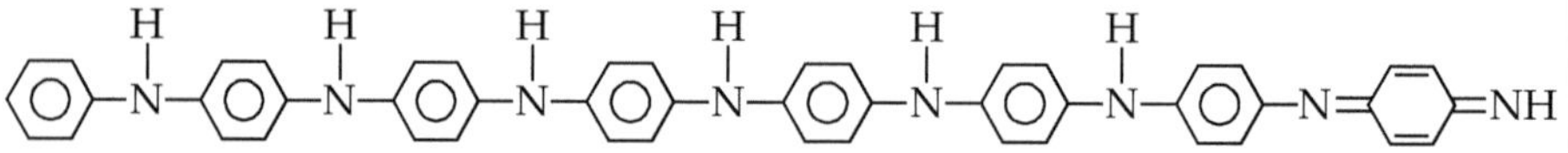

Violet substance, forming yellowish–green salts. Soluble in 80% acetic acid with a grass–green colour.

▶ **Emeraldine (diquinonoid state)**

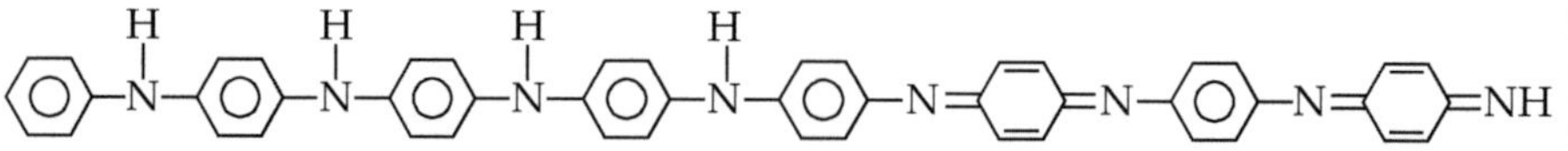

Violet blue substance, forming green salts. Soluble in 80% acetic acid, concentrated H_2SO_4 and 60% formic acid. The salts are stable but the base is slowly oxidized by air to nigraniline base.

▶ **Nigraniline**

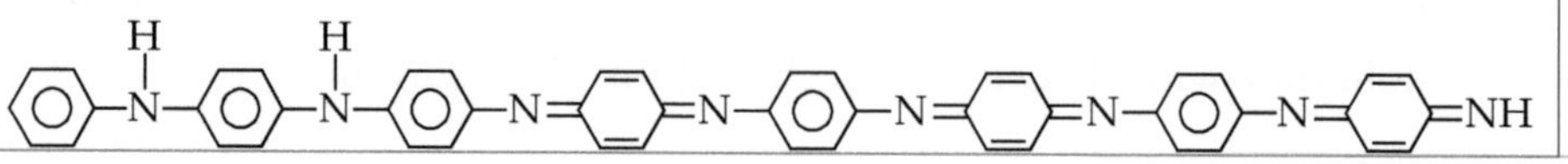

Dark blue substance, forming blue salts. Soluble in 80% acetic acid, pyridine, 60 % formic acid and concentrated sulfuric acid. The base is stable but the salts are very unstable.

▶ **Pernigraniline (tetrequinonoid structure)**

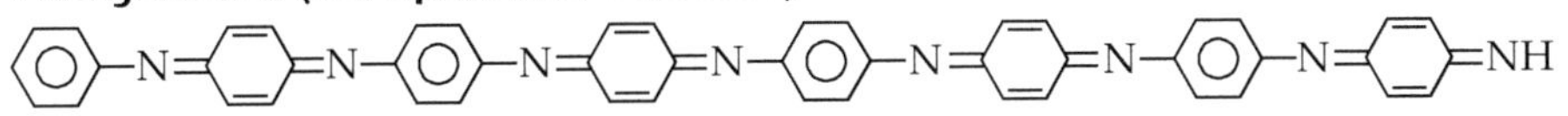

Purple substance, forming purple salts. Soluble in 80% acetic acid with a violet color. unstable, quickly undergoing degradation to the lower quinonoid stages.

▶ **Aniline octamers :** Structures and physicochemical properties between all five boundary oxidation states. The top row is called leucoemeraldine (LEB) oxidation state which is fully reduced state of PANI (amine content 100%). The second row shows the protoemeraldine oxidation state *i.e.* 25% oxidized form (amine content 75% and immine content 25%) of PANI. The third row is the half-oxidized emeraldine (EB) form of PANI. The fourth row shows the 75% oxidized state known as nigraaniline (amine content 25% and immine content 75%) and the last row is fully oxidized form (pernigraanilne). The immine nitrogen atoms can be protonated in whole or in part to give the corresponding salt, the degree of protonation of the polymer chain depending on its oxidation states and on the pH of the aqueous acid. See the following figure for what happen to a polymer chain when it is treated with a solution of different pH. At pH ≥ 10, the structure is fully reduced leucoemeraldine structure. If the pH value is changed to a value between 2 and 5, the LEB structure is changed via a protoemeraldine and/or emeraldine structure. If the pH values are lowered further to less than 2, the PANI chains become protonated or oxidized.

pH ≥ 10

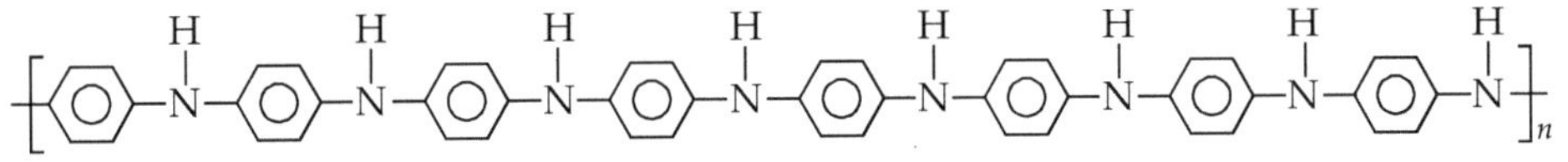

2 ≥ pH ≤ 5

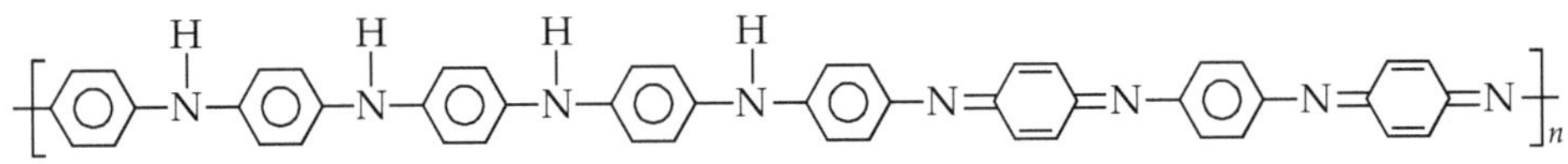

pH < 2

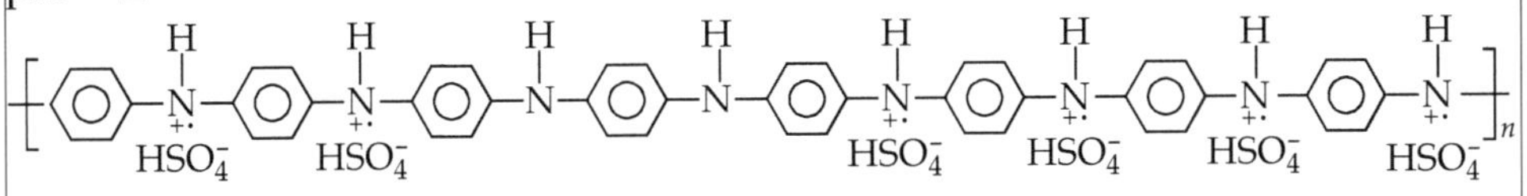

and/or

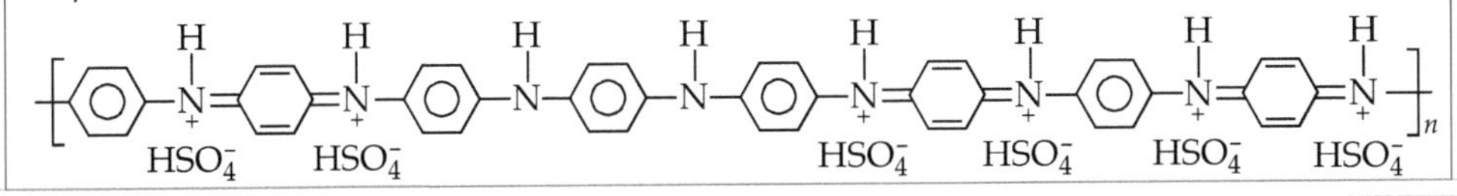

and/or

$$\left[\cdots\right]_n$$

and/or

$$\left[\cdots\right]_n$$

Polaron / bipolaron form of PANI in H_2SO_4 medium.

To compensate the positive charges from the protons attached to the nitrogen atoms, counter ions are absorbed by PANI chains. The protonation step is associated with the formation of radical cations or polarons and/or divalent cations or bipolarons. Subsequently during the 1980's MacDiarmid *et al.* showed that polyaniline can be prepared in the four idealized forms, at least when its synthesis and subsequent treatment are carried out in aqueous media. The polymer in the left hand column are considered as being derived from the amine, 1A (A = amine) and hence designated (in increasing degree of oxidation) as 1A and 2A. The materials in the right hand column are regarded as being derived from the ammonium salt, 1S (S = salt) and are hence designated (in increasing degree of oxidation) as 1S and 2S. All four forms are inter-convertible either by oxidation/reduction or by treatment with acid/alkali. Protoemeraldine, emeraldine and nigraniline are the intermediate oxidation states of the above mentioned forms of PANI.

1A
Insulating

1S
Insulating

Reduce / Oxidize

Reduce / Oxidize

2A
Insulating

2S
Metalic

The four idealized forms of polyaniline :

In order to understand the chemical nature of polyaniline, it is necessary to correlate both the colour and the conductivity with structure in a way that is compatible with the conditions of preparation (oxidative polymerization) and the circumstances in which inter-conversion between various forms takes place. With the present state of knowledge of the nature of the various forms of PANI and their transformations, Stejskal *et al.* used the following scheme.

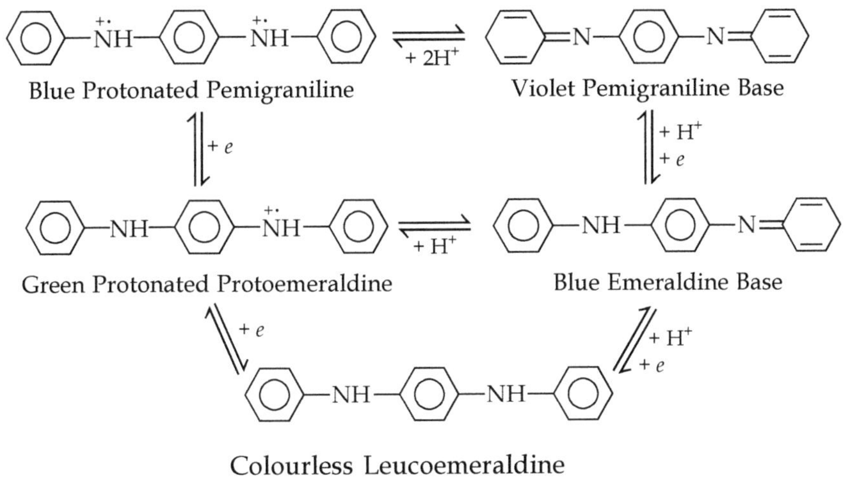

Blue Protonated Pemigraniline

Violet Pemigraniline Base

Green Protonated Protoemeraldine

Blue Emeraldine Base

Colourless Leucoemeraldine

Polyaniline forms and their inter-conversions.

9.15 Molecular Weight of Polyaniline

There are lots of reports in literature about the chemical and electrochemical synthesis of PANI (described in earlier section) but very few reports have in literature about its molecular weight. Molecular weights of PANI by chemical methods are much higher than electrochemically synthesized polymer. Electrochemical polymerization of aniline involves high oxidation potential (up to 0.9 V vs SCE) while chemical polymerization has potential of about 0.36 V. It has been seen that potentials higher than 0.6 V vs SCE induces degradation of PANI. The lower molecular weight for PANI using electrochemical methods is attributed to the higher oxidation potential used during polymerization. MacDiarmid *et al.* showed that the weight average molecular weight (M_w) usually varies from 1000 to 50,000 depending on the condition used for electro-polymerization and/or on the method of molecular weight determination. MacDiarmid *et al.* have observed the molecular weights (M_n = 25,000 and M_w = 64,000) for PANI by GPC (Gel Permeation Chromatography), where as another group led by Adams et al. have measured M_n values of only 2500 – 6000 for PANI by end group analysis, though in both the cases PANI preparation condition is same. So there is a large discrepancy in the estimation of molecular weight of PANI. Adams *et al.* is accounted this discrepancy as a result of the problem of effectively disrupting the strong intermolecular interactions, when dissolving undoped PANI in its standard solvent, NMP. As PANI.HCl is insoluble in common solvents, almost all the studies on molecular weight of PANI are carried out using the undoped form (EB). The major problem for studies of molecular weight of undoped PANI is that, it is not completely soluble in NMP. The viscosity average molecular weight (M_v) of PANI (EB) may be estimated by measuring the viscosity of EB solution in H_2SO_4 (97%). Since the K and α values ($[\eta] = KM^{\alpha}$) for polyaniline in H_2SO_4 are not known, then by this approximation of taking the K and ± values of rigid chain poly – p(phenyleneterephthalamide) [$k = 1.95 \times 10^{-6}$ and $\alpha = 1.36$] the M_v of PANI (EB) is 12000 – 15000 and for flexible chain viz. 40,0000. Adams *et al.* compared the molecular weight obtained by solid state ^{15}N NMR and GPC data (in NMP).

9.16 Optical and Electronic Properties of Polyaniline

There exists a single broad polaron band deep in the band gap of conducting ES form of PANI. This band is half filled, giving rise to an ESR signal. These band structure calculations are in agreement with the observed UV-vis-NIR spectra of emeraldine salts (ES), PANI. HA. In their "compact coil" conformation, emeraldine salts typically exhibit three peaks: one π-π^* (band gap) band at ca. 330 nm and two visible region bands at ca. 800 and 430 nm that may be assigned as π to polaron band and polaron to π^* band transitions, respectively.

Electronic band structures have also been calculated for each of the base forms of polyaniline, namely, the fully reduced leucoemeraldine base (LEB), the half-oxidized EB and the fully oxidized pemigraniline base (PB). The observed UV-visible spectra of LEB, EB and PB are in good agreement with these calculated band structures. The lowest energy absorption band for LEB occurs at ca. 320 nm and may be assigned to the π-π^* electronic transition, *i.e.* between the valence and conduction bands. For EB, as well as a similar low wavelength π-π^* band, there is a strong band at ca. 600 nm that has been attributed to a local charge transfer between a quinoid ring and the adjacent imine-phenyl-amine units giving rise to an intramolecular charge transfer exciton. Pemigraniline base also exhibits two absorption peaks: a π-π^* band at ca. 320 nm and a band at ca. 530 nm assigned to a Peierls gap transition. Protonation of PB causes a violet-to-blue color change due to the formation of pemigraniline salt (PS). This change is associated with the loss of the PB band at 530 nm and the appearance of a strong PS peak at ca. 700 nm. Cao *et al.* have reported three spectral features at 1 eV, 1.5 eV and 3 eV for a polyaniline film spun cast from a sulphuric acid solution onto sapphire substrates. (1) Intrachain absorption at 1 eV and 3 eV. (2) Interchain absorption at 1.5 eV. The authors have reported spectra of leucoemeraldine and emeraldine base. Leucomeraldine, which is an insulator with large band gap, showed the onset of a π-π^* transition around 3 eV with a peak at 3.7-3.8 eV. The emeraldine base gave two principal absorption bands having maxima at 2 eV and 3.9 eV. The over oxidized polyaniline showed an intense absorption band at 2.2 eV and a well-defined peak near 4 eV.

9.17 Vibrational Spectra of Polyaniline

PANI has different oxidation states. For this it shows insulator to metal transition during sweeping of electrochemical potential and changes in the pH of the medium. Electrochemical and chemical doping yield different structures which can be ascertained using in-situ FTIR spectroscopy coupled with cyclic voltammetric studies. Infrared spectroscopy is a powerful tool to determine the structural changes that occur during doping or dedoping process. Various groups have reported FTIR results of polyaniline. It has been shown that the absorption frequencies are strongly influenced by the electrochemical potential. The intensity of the 1570-1590 cm^{-1} band relative to 1500 cm^{-1} is a measure of the degree of oxidation of the polymer film. Similarly, when going from the reduced state (-200 mV) to the oxidized state (400 mV) a new peak at 1375 cm^{-1} is observed, whose intensity increases with the degree of oxidation. The maximum intensity due to counter ions is observed at 400-600 mV. During electrochemical doping the absorption intensities increase at 1563, 1477, 1300, 1244, 1158, and 1041 cm^{-1} while absorption at 1507 cm^{-1} show a decrease in intensity. When the anodic potential exceeds 600 mV increases in absorption intensities at 1629, 15781, 1507, 1376, 1339 and 1100 cm^{-1} are observed. When PANI is held at 1 V, the absorption intensities of 1654, 1451, 1313, 1080, 944 and 885 cm^{-1} increase and 1624, 1123 cm^{-1} decreases. During oxidation, absorption peak of benzenoid at 1500 cm^{-1} decrease while for quinoid rings at 1560-70 cm^{-1} increase. The appearance of a band at 1375 cm^{-1} during oxidation is due to a semiquinoid ring N-ring mode and fully quinoid ring absorption is observed at 1630 cm^{-1} and 1150 cm^{-1}. PANI also has a characteristic peak at 1160 cm^{-1}, corresponding to a vibration mode of quinonoid structure (N=Q=N) of PANI (EB) with partial electronic-like character. In over oxidation the decrease in intensity at 1630 cm^{-1} indicates degradation of the quinoid structure to give benzoquinone as indicated by absorption peaks at 1654, 1313, 1080, 944 and 885 cm^{-1}. Neugebauer *et al.* have studied the FTIR spectrum of polyaniline by carrying out FTIR measurements during cyclic voltammetric experiments at pH 4.5. These authors observed a sharp increase in absorption at 1580, 1490, 1388, 1330, 1245, and 1140 cm^{-1} and a decrease at 1600, 1510 and 1300 cm^{-1} during the anodic process (oxidation). These changes may be due to the transition of benzenoid (1600, 1510 cm^{-1}) to quinoid (1580, 1490 cm^{-1}) that is similar to the oxidation of polyaniline under acidic conditions. The peaks observed in the IR spectra of polyaniline are given below

$C{-}N$ stretching and $C{-}C$	Stretching	1288-1214
$C{-}N^+$ stretching and $C{-}C$	Stretching	1228-1221
$C{=}N$ stretching and $C{-}C$	Stretching	1522-1480
$C{-}N^+$ stretching and $C{-}C$	Stretching	1401-1383
	Benzenoid (B)	**Quinoid (Q)**
Aromatic ring	1625-1615	1580-1590
	840-820	780-810
CH bending (in plane)	1185-1175	1160-1180
CH (out of plane)	885-870	
NH stretching	3400-3100	
NH_2^+	3000-2450	
($N{=}Q{=}N$) (electronic-like character)		1150-1170
	Benzenoid	**Quinoid**
CH (out of plane)	822-810	780-800
Ring stretching	1540-1495	1540-1570
CN stretching and CH bending	1315-1285	1290-1300

* Values are in cm^{-1}

9.18 Thermal Properties of Polyaniline

Thermal stability of PANI has been studied by many workers using thermogravimetric analysis (TGA), differential scanning calorimetry (DSC) and infrared spectroscopy. The thermal response of PANI can be roughly divided into three major steps: (i) the loss of moisture, (ii) removal of free HCl and unreacted monomer and the elimination of dopant and (iii) the destruction of skeletal backbone at increasing temperature. LaCroix and Diaz investigated thermal stability of electrochemically synthesized polyaniline by TGA under helium atmosphere. Two major weight losses of the acid doped PANI, one at 60-80°C and other at 185-195°C were observed, which were attributed respectively to the loss of moisture and the loss of HCl. Hagiwara *et al.* reported that the conductivity of chemically synthesized polyaniline hydrochloride (PANI.HCl) deteriorates upon thermal aging at 150°C in air and in vacuum due to the elimination of HCl from the amino group and the simultaneous chlorination of the aromatic ring in the polymer. Wei *et al.* showed that three major weight losses of HCl doped PANI occurs at around 100, 200 and 500°C in nitrogen and air but for the undoped PANI weight loss occurs at 500°C. So, the weight loss around 500°C could be attributed to the structural decomposition of the polymer. But, there is slightly different type report for thermal characteristic of chemically synthesized PANI base form. Ding et al. showed from the TGA study that there are two major type weight losses for the polyaniline (EB) powder, regardless of the experimental environments. The first weight loss at the lower temperature at around 100°C results from moisture evaporation and perhaps out gassing of unknown small molecules. The second weight loss at the higher temperature indicates a structural decomposition of the polymer as mentioned by other workers. In air atmosphere, polyaniline (EB) powder started to degrade at 370°C and was completely decomposed at 600°C. There was no residual powder left in the sample pan after the test. The major weight loss of only 14.23% in the nitrogen atmosphere was observed in the range of 475-600°C. Ding et al. observed an endothermic peak at 50-140°C and an exothermic peak at 185-350°C for PANI (EB) when measured in DSC. They attributed that the first peak is due to the removal of water and the exothermic peak has been related to cross-linking reaction but not due to decomposition of EB since there was no weight loss observed in this temperature range by TGA study. To further understand the possible thermal transitions which may be hidden in the DSC peaks, Ding *et al.* used MDSC (modulated DSC) to study EB powder. MDSC is a high performance version of traditional DSC that not only records the total heat flow (same as DSC), but also identifies the reversing and non-reversing heat flow, respectively. From this result Ding et al. found two glass transitions (T_g) at ~ 70°C and ~250°C respectively from reversible heat flow of EB powder measured by MDSC. This is different from conventional DSC since they are swamped by non-reversing exothermic and endothermic peak. The lower T_g at ~ 70°C was measured before crosslinking of the EB powder. The higher T_g at ~ 250°C has been attributed to the partially crosslinked polyaniline.

9.19 Mechanical Properties of Polyaniline

The mechanical properties of PANI depend on its structure and the morphology. The electropolymerized PANI film is highly porous and consequently has low mechanical strength whereas the polymer made from solution casting is less porous and have higher mechanical strength.

Kitani *et al.* have electrochemically prepared freestanding films of PANI in the reduced (leucoemeraldine) state. When oxidized to the emeraldine state, the films become brittle. This change in mechanical properties is due to the increased interchain bonding between charged chains which result in a decrease in toughness. The polymerization potential has also been found to influence the mechanical properties of PANI films. The most extensible films were formed at a polymerization potential of 0.65 V (vs. Ag/Ag$^+$), which displayed an extension to break of around 40%. Preparation of the PANI films at 0.8 V and 1.0 V resulted in more-brittle films. It was suggested that degradation of the PANI at polymerization potentials in excess of 0.8V might explain the poor properties of the 1.0 V film. The difference in behaviour of the films prepared at 0.65 V and 0.8 V was attributed to differences in their crosslink density. Unfortunately, structural characterizations of the PANI films prepared under these conditions were not conducted, making it impossible to explore the structure-property relationships in more detail.

For solution cast films, researchers have shown that high molecular weight PANI is the main criteria to successfully prepare fibers and films with adequate mechanical properties. Laughlin and Monkman have shown that a molecular weight of 130,000 g/mol is sufficient, whereas Mattes *et al.* have investigated the effect of molecular weight in the range 100,000 to 300,000 g/mol. The higher the molecular weight, the more the films could be drawn and the higher was the tensile strength. As expected, the stretching of PANI films and fibers has a dramatic effect on the mechanical properties. EB films can be stretched by more than five times their original length (at 140°C) and this process has been shown to produce a 10-fold increase in tensile strength (to 226 Mpa). Similarly, thermal stretch orientation of EB fibers produces a tensile strength of 318 MPa which reduces to 150 MPa when HCl doped. Similar increase in elastic modulus would be expected to result from the mechanical drawing, since the process causes considerable alignment of the polymer chains in the draw direction. Such mechanical orientation also increases the crystallinity and conductivity of the PANI films and fibers. Polyaniline can be successfully plasticized to improve the ductility and toughness. Residual solvent (NMP or high boiling solvent) in solution-cast films and fibers undoubtedly affects the mechanical properties: increasing elongation at break and reducing the elastic modulus. Fedorko et al. have recently shown dramatic improvements in the ductility, of PANI (ES) films using plasticizing dopants. Thus, the di(2-ethylhexyl)ester of 4-sulfophthalic acid (DEHEPSA) was used to prepare PANI (ES) that were soluble (in the ES form) in dichloroacetic acid. Cast films had similar conductivities to ES-CSA of ~100 S/cm. Tensile testing gave an elongation at break of 28% for the ES-DEHEPSA films compared with only 2% for the ES-CSA. The tensile strength for both films were similar 14 and 16 MPa respectively. This tensile strength is comparable with that obtained for un-oriented EB films. Most fibers and films of PANI have been prepared from a solution of emeraldine base and converted to the emeraldine salt by acid doping. The choice of dopant acid has a profound effect on mechanical properties. MacDiarmid et. a1. have shown that the mechanical properties depend in a complex way on dopant, casting solvent and polymer molecular weight. Full details of the effects of polymer structure (as influenced by dopant and solvent) on the mechanical properties are yet to be elucidated. Thermal analysis has been used to study thermal transitions in polyaniline, specifically the glass transition.

This polymer has been shown to display a distinct T_g with the transition being sensitive to the degree of plasticization. DMA studies on PANI (EB) films cast from NMP have shown the T_g to shift from 158°C to 99°C as the NMP content decreases from 18% to 10% (w/w). Gregory has evaluated the thermal properties of the LB form of PANI using Differential Scanning Calorimetry (DSC) and DMA. A broad endotherm was observed at ~385°C substantially higher in temperature than the T_g at ~200°C. The endotherm was attributed to the melting of a crystalline phase as supported by XRD and microscopy data. DMA data also confirmed significant softening of the polymer at ~375°C. The fibers and films used by Gregory were prepared from DMPU, a solvent that inhibits gelation of the EB in solution.

9.20 Application of polyaniline based material

Conducting polymers has wide applicability in modern material science. Many institutes and commercial establishments involve multidisciplinary research into different types of chemical synthesis, polymer processing, electronics, physics and applied physics. The combination of metal-like or semiconducting behaviours and processing of classical polymers has created opportunities for scientists and technologists to investigate possible technological applications. Several reports and review papers have indicated promising applications of polyaniline and at least a hundred companies are involved in the test production of conducing polymers. Polyaniline is applicable in Conductivity (plastic Batteries), Photoconducting (LEDs, photocopiers), piezoelectric micromotors transducers, photochemical reaction (optical storage, lithography), conducting composite (super capacitors), solid state sensors, membranes (gases separation), nonlinear optics (harmonic generators), electrochromic display devices, ferromagnetism (magnetic recording) and conductive surface (EMI/ESD).

1. **Write down the general structure of polyalkenes, styrene copolymers, Related polymers of poly(vinyl chloride), Related polymers of poly(vinyl acetate), acrylic polymers, fluoropolymers, polyamides, polyurethanes, silicone polymers, polydienes and polycarbonates.**

Ans. The general structure of polyalkenes, styrene copolymers, Related polymers of poly(vinyl chloride), Related polymers of poly(vinyl acetate), acrylic polymers, fluoropolymers, polyamides, polyurethanes, silicone polymers, polydienes and polycarbonates are as follows

General structure of polyalkenes 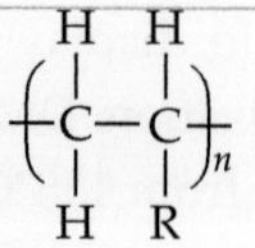

General structure of styrene copolymers 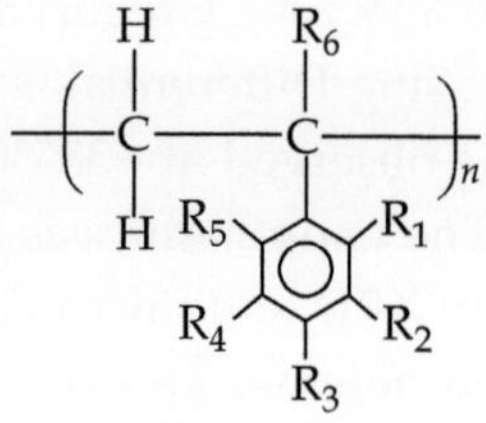

General structure for Related polymers of poly(vinyl chloride) 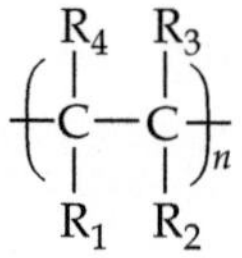
[R_1, R_2, R_3, R_4 = H, F, Cl, Br but one component must be Cl]

General structure for Related polymers of poly(vinyl acetate) 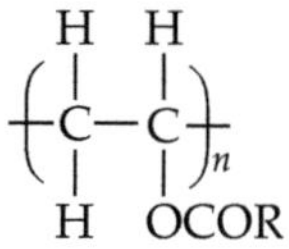

General structure of acrylic polymers 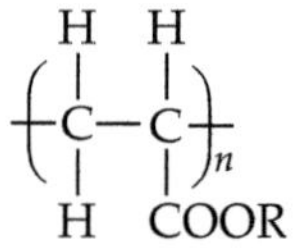

General structure of fluoropolymers 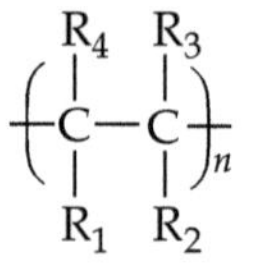

[R_1, R_2, R_3, R_4 = H, F, Cl, Br but one component must be F]

General structure polyamides 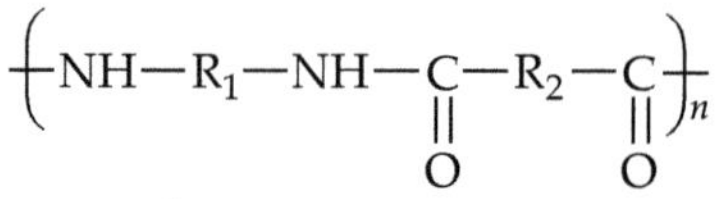

General structure of polyurethanes 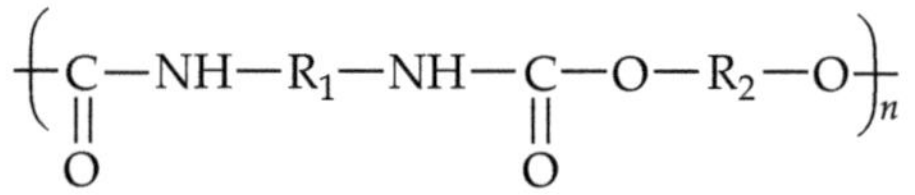

General structure of silicone polymers 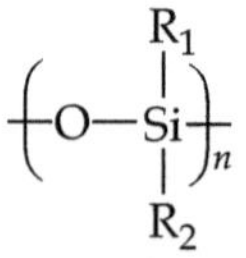

General structure of polydienes 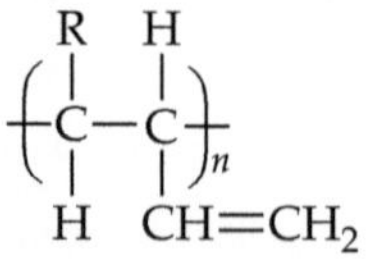 and 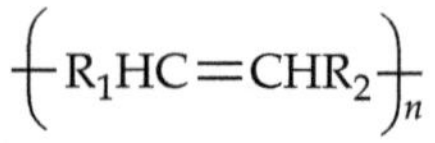

General structure of polycarbonates 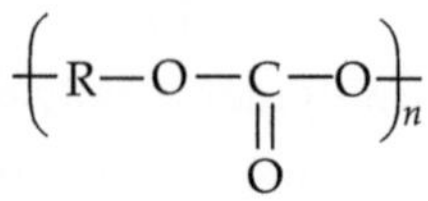

2. Write down the structure of novalac and bakelite.

Ans. The structure of novolac is

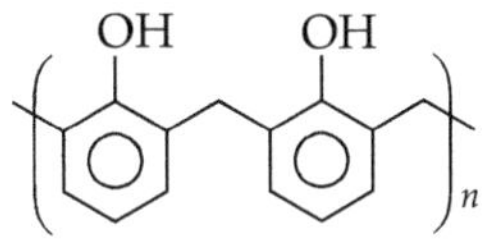

(structure of novolac)

The structure of bakelite is

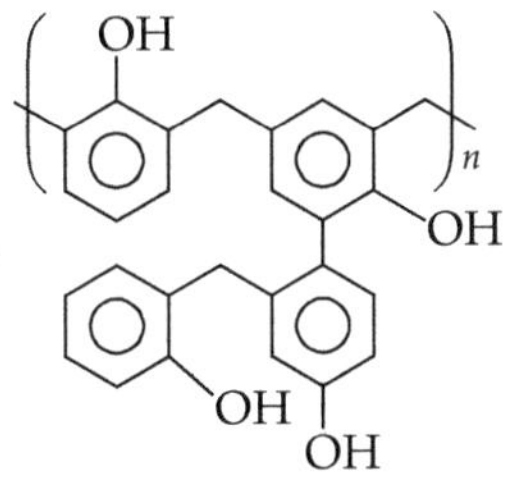

(structure of bakelite)

3. Write down the structure of polyacetylene, polyaniline, poly(*p*-phenylenesulphide), polypyrrole, polythiophene, polyphenylene and Poly(*p*-phenylene vinylene).

Ans. The structure of polyacetylene, polyaniline, poly(*p*-phenylenesulphide), polypyrrole, polythiophene, polyphenylene and poly(*p*-phenylene vinylene) are shown below.

Polyacetylene *(structure)*

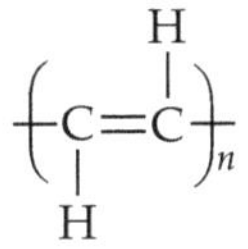

Polyaniline *(structure)*

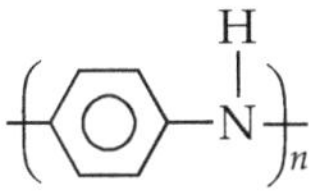

Poly(*p*-phenylenesulphide) *(structure)*

Polypyrrole *(structure)*

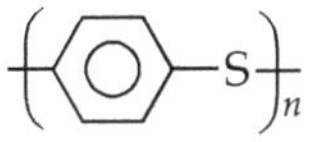

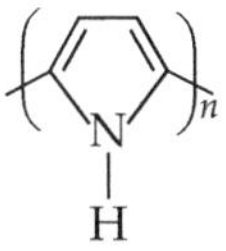

Polythiophene *(structure)*

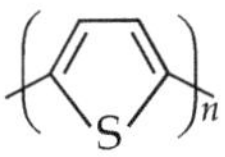

Polyphenylene *(structure)*

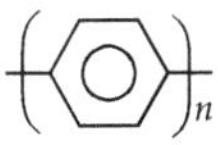

Poly(*p*-phenylene vinylene) *(structure)*

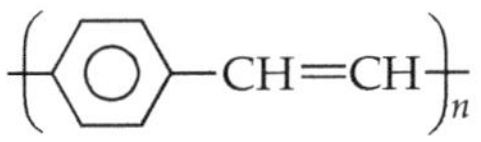

4. Write down the application of polycarbonates.

Ans. Polycarbonatesare widely used in construction materials (safety helmets, power tools, appliance housings, glazing for windows, doors and face shields), medical equipment, mobile phone, compact discs, sun glasses, automotive,aircraft, and security components.

5. Write the properties of polydienes.

Ans. polydieneshave very excellent elastic properties, low glass transition temperature and good UV stability. It is chemical resistance (mainly oxygen and ozone), low temperatures resistance. Sometimes polydienes contain unsaturated sites which can furtherbe vulcanized to produce an elastomeric network. Styrene-butadiene rubber, polybutadiene, nitrile-butadiene rubber, polychloroprene and polyisoprene are the common diene elastomers which have excellent properties for many applications.

6. Where are the polyurethane polymers are used.

Ans. Polyurethanes are the top classes of polymers with excellent properties which make them in a wide range of applications. They are applicable in producing flexible and rigid foams like mattresses, carpet, cushions and armrests, insulations for commercial and residential buildings, tanks, pipes, water heaters, refrigerators, and freezers. Polyurethanes are also applicable as elastomeric products like bumpers, fenders, steering wheels, instrument and door panels, and gaskets for trunks, windows and windshields. They are also applicable in adhesives, coatings, sealants, and synthetic fibers.

7. How fluoropolymers are prepared?

Ans. Fluoropolymers are prepared by conventional free-radical polymerization method. But anionic ring opening polymerisation or UV-catalysed radical oxopolymerisation processes are also used for its synthesis.

Exercise

A Multiple Choice Type Questions(MCQ) (Each Question carries 1 mark)

1. What is the simplest synthetic polymer?

    a. Polyethylene   **b.** Polyaniline

    c. PMMA   **d.** PNIPAM

2. PVC is

    a. synthetic plastic polymer   **b.** natural polymer

    c. rubber   **d.** both b and c

3. Which is/are not polymer?

    a. PVC   **b.** DNA

    c. HDPE   **d.** sucrose, glucose and fructose

4. polyolefin is

    a. toxic   **b.** non-toxic   **c.** highly toxic   **d.** explosive

5. Which is the safest plastic?

    a. ULDPE   **b.** LDPE

    c. HDPE   **d.** VLDPE

6. Which plastic is toxic?

    a. Polyvinyl chloride   **b.** PE

    c. Polypropylene   **d.** None of these

7. Is scratched plastic safe?

    a. No   **b.** Yes

    c. Do not depend   **d.** None of these

8. Which solvents can dissolve polypropylene and polyethylene at room temperature?

    a. benzene   **b.** 1,2,4-trichlorobenzene

    c. water   **d.** ethanol

9. Why are olefins (alkenes) good monomers for polymerization reactions?

    a. the weak π-bonds form strong σ bonds to other monomer units

    b. easily available

    c. non-reactive

    d. not hazardous

10. How can chemists? control which type of polyethylene (LDPE vs. HDPE) is generated?

    a. Only reaction condition   **b.** Only catalyst

    c. Proper choice of catalysts and reaction conditions

    d. None of these

11. Which is a synthetic aromatic polymer

    a. Polystyrene   **b.** Only catalyst

    c. Only reaction condition   **d.** None of these

12. Smoke detector housings is made by license plate frames

    a. polypropylene   **b.** Polyaniline

    c. Polyethylene   **d.** Polystyrene

13. License plate frames is made by

    a. Polystyrene   **b.** Polyaniline

    c. Polyethylene   **d.** Polypropylene

14. Is polystyrene foam safe?

    a. yes   **b.** no

    c. not important   **d.** None of these

15. Polystyrene utensil is used for

    a. cooking   **b.** reheating

    c. depend on reactivity with food   **d.** None of these

16. Which of the following is used as a catalyst for the alkylation of benzene in the process of formation of styrene?

a. $TiCl_4$

b. MgO

c. FeO

d. $AlCl_3$

17. Which of the following can stabilize the vinyl benzene?

a. *t*-butyl catechol

b. *t*-butyl hexanol

c. lactone

d. none of these

18. How many stages do the bulk polymerization of styrene passes through?

a. 3

b. 1

c. 2

d. 4

19. What kind of polymer is the commercially used polystyrene?

a. atactic

b. isotactic

c. syndiotactic

d. all of these

20. What is the percentage value of elongation at break for polystyrene?

a. 200-240%

b. 100-300%

c. 20-130%

d. 1-3%

21. What is the approximate density of polystyrene in g/cc?

a. 1.054

b. 0.96

c. 0.915

d. 1.5

22. Which of the following property of polystyrene is improved, when traces of unreacted monomers are removed?

a. brittleness

b. melting point

c. crystallinity

d. none of these

23. Which of the following grade of polystyrene is useful in optical applications?

a. high impact (HIPS)

b. general-purpose (GP-PS)

c. expanded grades

d. none of these

24. Which of the following grades of poly vinyl chloride is unsuitable for insulation purposes?

a. suspension grade

b. bulk grade

c. emulsion grade

d. None of these

25. Which of the following solvent cannot solubilize polyvinyl chloride into it?

a. hexane

b. cyclohexanone

c. tetra hydrofuran

d. nitrobenzene

26. Which of the following monomer pairs forms a copolymer that exhibits excellent chemical resistance and self-extinguishing properties?

a. vinyl chloride + vinylidene chloride

b. vinyl chloride + vinyl acetate

c. vinyl chloride + acrylonitrile

d. none of the mentioned

27. Which of the following compound is used in fire retardant compounds of PVC?

    a. dibutyl phthalate   **b.** tricresyl phosphate

    c. barium phenate   **d.** dibutyl sebacate

28. Which of the following compounds can process the flexibility applications of PVC?

    a. stabilizers   **b.** lubricants

    c. plasticizers   **d.** colouring matters

29. Which is the most widely used polymerization technique to produce polyvinyl actate?

    a. emulsion polymerization   **b.** solution polymerization

    c. bulk polymerization   **d.** all of the mentioned

30. In which of the following solvents, polyvinyl acetate is immiscible?

    a. ethanol   **b.** methanol

    c. benzene   **d.** acetone

31. Which of the following pair leads to the formation of polyvinyl acetals?

    a. Polyvinyl acetate + ester   **b.** Polyvinyl alcohol + esters

    c. Polyvinyl chloride + aldehydes   **d.** Polyvinyl alcohol + aldehydes

32. What happens when the degree of alcoholysis of polyvinyl alcohol is increased beyond 90%?

    a. water solubility decreases   **b.** water solubility increases

    c. decolouration occurs   **d.** water insolubility

33. Which of the following is used as a catalyst in the formation of polyvinyl butyral?

    a. H_2SO_4 **b.** Na_2SO_4

    c. CrO_3 **d.** $(CH_3COO)_2Cd$

34. What is the correct structure of monomer of poly methyl methacrylate?

    a. $CH_2{=}C(CH_3){-}CO{-}O{-}CH_3$ **b.** $CH_3{-}CO{-}O{-}CH_3$

    c. $CH_3{-}CO{-}O{-}CH{=}CH_2$ **d.** $CH_3{-}CO{-}O{-}C(CH_3){=}CH_2$

35. Which of the following mechanism is used in the formation of commercially used polymethyl methacrylate?

    a. cationic mechanism   **b.** free radical mechanism

    c. anionic mechanism   **d.** cannot be said

36. What is the specialized application of PMMA, keeping in mind its excellent optical clarity and high atmospheric resistance?

    a. hard contact lenses   **b.** construction

    c. decoration of appliances   **d.** insulation wires

37. Which of the following polyacrylate is used in paints and adhesives formulation?

    a. poly (methyl acrylate)   **b.** poly (2-ethylhexyl acrylate)

    c. poly (butyl acrylate)   **d.** poly (ethyl acrylate)

38. How is the polymer formed by polymerization of acrylonitrile obtained in the form of?

a. colloids

b. suspension

c. homogeneous solution

d. precipitate

39. What is the melting point of polymer formed by polymerization of tetrafluoroethylene?

a. 327°C   **b.** 156°C   **c.** 100°C   **d.** 533°C

40. How is the final product of polymerization of tetrafluoroethylene obtained in the form of?

a. granular or aqueous dispersion   **b.** granular

c. granular or homogeneous solution   **d.** suspension

41. Which of the following is false about the properties of polytetrafluoroethylene?

a. high crystallinity   **b.** low dissipation factor

c. low chemical inertness   **d.** low friction coefficient

42. Which of the following fluoro polymer film is sold under the name of Tedlar by DuPont?

a. hexafluoropropylene   **b.** polytetrafluoroethylene

c. polyvinylidene fluoride   **d.** polyvinyl fluoride

43. What is the approximate density of the polymer polytetrafluoroethylene?

a. 1.8 g/cm^3 **b.** 2.2 g/cm^3

c. 0.6 g/cm^3 **d.** 1.5 g/cm^3

44. Name the polymer formed by the polymerization of ω-amino caproic acid using water as a catalyst?

a. nylon 6   **b.** nylon 6,6   **c.** nomex   **d.** abs

45. Which of the following mechanism takes place for the formation of high molecular weight Nylon 6 from caprolactum?

a. anionic mechanism   **b.** cationic mechanism

c. radical mechanism   **d.** none of the mentioned

46. Which of the following polyamides are known as aramids?

a. aliphatic polyamides   **b.** aromatic polyamides

c. unsaturated polyamides   **d.** all of the mentioned

47. Polyamides have a limited use in the areas where long exposure to temperatures above 70-90°C is required, due to their tendency of surface oxidation in air. Which heat stabilizer can permit their use even in higher temperatures?

a. copper salts   **b.** sodium salts

c. potassium salts   **d.** none of the mentioned

48. Which polyamide is formed when polymerization of isophthaloyl chloride and *m*-phenylene diamine takes place?

a. kevlar

b. nomex

c. nylon 6

d. SBR

49. Which of the following monomer pair polymerizes to give an aramide?

a. terephthaloyl chloride + phenylene diamine

b. hexamethylene diamine + sebacic acid

c. hexamethylene diamine + adipic acid

d. none of the mentioned

50. Which polyamide can be used as substitute of steel in belted radial tires?

a. nylon 66

b. nomex

c. nylon 6

d. kevlar

51. How many methylol groups are present in a typical novolac molecule?

a. 0   **b.** 6-8   **c.** 3-5   **d.** 2

52. What is the name of the B- staged resin formed in resinification process?

a. resitol

b. resol

c. resite

d. novolac

53. Which of the following phenolic resins are suitable for the decorative laminates?

a. resites

b. ammonia catalyzed resols

c. spirit resols

d. caustic soda catalyzed resols

54. What is the approximate phenol formaldehyde ratio used is the preparation of casting grade phenolic resins?

a. 1:2.2   **b.** 2:1   **c.** 3:1   **d.** 1:2

55. What type of phenolics is extensively used as objects of art and jewellery?

a. laminated phenolics

b. cast phenolics

c. oil-soluble phenolics

d. resols

56. How is the cross-linking of novolacs better accomplished along with the presence of additional formaldehyde?

a. use of basic catalyst

b. use of acidic catalyst

c. additional phenol

d. none of the mentioned

57. Which of the following properties of polyurethane rubbers is incorrect as compared to other rubbers?

a. excellent abrasion resistance

b. excellent resistance to acids

c. excellent resistance to oxygen

d. higher tensile strength

58. Where is the polyurethane foams extensively used as?

a. paint rollers

b. adhesive

c. printing rollers

d. fabric coatings

59. Which of the following monomer pairs polymerize to give epoxy resin?

    a. epichlorohydrin and bisphenol A   **b.** bisphenol A and diphenyl carbonate

    c. bisphenol A and phosgene   **d.** none of the mentioned

60. What is the common name of poly (ethylene terephthalate)?

    a. viton   **b.** technora   **c.** dacron   **d.** nomex

61. What is the common or brand name of the polymer polyacrylonitrile?

    a. mylar   **b.** orlon   **c.** technora   **d.** ultem

62. Silicones are called inorganic polymers due to absence of which atom in the main backbone chain.

    a. Carbon atom   **b.** Oxygen atom

    c. Nitrogen atom   **d.** Hydrogen atom

63. The catalyst used in the preparation of poly-siloxanes is

    a. Lewis acid   **b.** Benzoyl peroxide

    c. Cumene hydroperoxide   **d.** Grignard reagent

64. Which of the following gives linear chain silicones on polymerisation by controlled hydrolysis?

    a. Dimethyl silicon chloride   **b.** Trimethyl silicon chloride

    c. Tetramethyl silicon chloride   **d.** Momethyl silicon chloride

65. On increasing the temperature, the viscosity of polysiloxanes

    a. Remains constant   **b.** Decreases

    c. Increases   **d.** First increases then decrease

66. natural rubber is a polymer of

    a. *trans* isoprene   **b.** *cis* isoprene

    c. Buna N   **d.** Buna S

67. Synthetic polymer that resembles to natural rubber is

    a. Neoprene   **b.** trans isoprene

    c. Buna N   **d.** Buna S

68. Which of the following is polycarbonate?

    a. Buna N   **b.** Acrilan

    c. Lexan   **d.** Buna S

69. Which of the following is used for making rechargeable batteries?

    a. Polyaniline   **b.** Polyester

    c. Polypyrrole   **d.** Polyacrylonitrile

70. The advantage of using conducting polymers in place metals is

    a. Light-weight   **b.** Cost

    c. Thermal conductivity   **d.** Solubility

71. During the polymerisation of polyaniline, the sequence of the colour change is

    a. Blue green >> green >> light blue >> copper tint

    b. Green >> copper tint >> blue green >> light blue

    c. Copper tint >> blue green >> green >> light blue

    d. Light blue >> blue green >> copper tint >> green

72. The color of emeraldine salt which is obtained is

    a. Copper tint   **b.** Blue green

    c. Green   **d.** Light blue

73. Which of the following is not an application of conducting polymers?

    a. Adhesives   **b.** Analytical sensors

    c. Electronics   **d.** Rechargeable batteries

74. Polyurethane rubber is also known as

    a. isocynate rubber   **b.** Thiokol

    c. neoprene   **d.** hypanol

75. Silicones contain alternate

    a. Si, C   **b.** Si, O

    c. C, H   **d.** C, N

76. Which of the following is conducting polymer?

    a. polyacetylene   **b.** HDPE

    c. LDPE   **d.** polycarbonate

B **Short Answer Type Questions (Each question carry either 2 or 3 marks)**

Write short notes on the following

1. Polyolefins

2. Polystyrene

3. Poly(vinyl chloride)

4. Poly(vinyl acetate)

5. Acrylic polymers

6. Fluoropolymers

7. Polyamides

8. Phenol formaldehyde resins

9. Polyurethanes

10. Silicone polymers

11. Polydienes

12. Polycarbonates

13. Polyacetylene

14. Polyaniline

15. Poly(*p*-phenylene sulphide)

16. Polypyrrole

17. Polythiophene

🖸 Long Answer Type Questions (Each question carry 5 marks)

1. Polymer melts are processed at high rates of strain, $\dfrac{d\gamma}{dt}$, in an extruder or injection.

    a) Sketch a typical plot of polymer viscosity versus rate of strain on a log-log plot showing the power-law region and the Newtonian plateau at low rates of strain.

    b) Explain using this plot why polymers are processed at high rates of strain.

    c) Using your own words propose a molecular model that could explain shear thinning in polymers.

    d) Sketch a plot of log viscosity versus log strain rate for a series of polymers of different molecular weight showing the effect of molecular weight on this plot.

2. The zero shear rate viscosity, η_0, displays power-law behaviour in molecular weight in two distinct and universal regimes.

    a) Sketch log viscosity versus log of molecular weight showing these two regimes and give the slope of the curve in these two regimes.

    b) Explain why two regimes are observed

    c) How does this plot relate to the definition of a plastic?

    d) Why would similar behavior be seen in viscosity versus concentration for high molecular weight polymer solutions?

3. a) What are the monomers used to synthesize nylon 6, 10?

    b) What solvents are used in the interfacial polymerization and in which solvent is each monomer dissolved?

    c) How is the polymer separated from the reaction mixture in this reaction?

    d) How is this reaction driven to a high extent of reaction, $p \geq 1$?

    e) Typically a step growth polymerization, such as the reaction of epoxy glue, proceeds through a low viscosity stage (application of the glue), a viscous stage (early stages of setting/tacky glue) and finally a solid stage (cured glue). The interfacial nylon synthesis does not display these stages. Is this really a step growth polymerization? Explain your answer.

4. The synthesis of glyptal involves a trifunctional and a bifunctional reactant in a condensation reaction.

    a) Give the monomers involved in glyptal in a stoichiometric chemical reaction equation. (Draw the structure and give the number of these monomers involved in a stoichiometrically balanced reaction.)

    b) Which solvent is used to make glyptal?

    c) Draw the structure of sodium acetate and explain why it was used in this reaction.

   d) Give a sequence of events that occur chronologically in the reaction to form Glyptal.

   e) Give the structure of glyptal and the structure of PET (PETE) and comment on the difference in properties between these two similar polyesters based on the difference in chemical structure.

5. a) Give the structure of PMMA and the structure of the polyacrylate used in the hydrogel. Compare the two and explain the differences in properties between PMMA and polyacrylate associated with this structural difference, i.e. why isn't PMMA a superabsorbent?

   b) What happens when uncrosslinked polyacrylate is added to water? Explain this.

   c) The swelling ratio Q for the hydrogel was about 100 to 500. Write an equation for the swelling ration as a function of the interaction parameter, χ, and the molecular weight between crosslinks, n.

   d) If you wanted a more robust gel what n would you choose.

   e) What happens to ? when salt is added to the hydrogel? Explain what happens to the hydrogel.

6. a) Draw the structure of PVA and boric acid.

   b) The combination of PVA and boric acid yields slime, but when NaOH is added a rubbery elastomer results. Explain this in terms of the percolation model for gelation.

   c) What happens to the molecular weight between crosslinks when NaOH is added? What happens when HCl is added?

   d) Gellation occurs at a fixed value of the extent of reaction, p_c, where c indicates the critical extent of reaction. A critical point is where some features of the system goes to infinity. What feature of a gelling system goes to infinity at p_c, the gel point? Make a rough plot.

   e) Give a function that relates the critical extent of reaction with the structure of a stoichiometric network and calculate p_c for the glyptal system using this function.

7. Polyurethane is formed from two liquids. After mixing the solution foams and expands fairly rapidly forming a solid foam after a few minutes.

   a) One of the liquids contains MDI. Give the full name and structure for MDI.

   b) What is the reactant (co-monomer) in the second liquid?

   c) Name a catalyst (give acronym) that might be in the second liquid.

   d) What role would water play if it were present in the second liquid?

   e) What happens if a diamine is used rather than what you have listed in part b?

8. a) Which two reactants were used to make the novolac?

   b) How do these reactants differ from those used to make a resole polymer?

   c) For the novolac what condition is needed?

   d) Outline the reaction scheme for formation of the novolac polymer.

   e) Why is the novolac procedure easier to demonstrate than the resole procedure?

9. For polyimides and epoxys

   a) Give the structure of an imide bond.

   b) Give the reactants that form a cyclic polyimide such as kapton.

   c) Show the two reaction steps to form a polyimide

   d) Give the structure of epichlorohydrin.

   e) Give the structure of a glycidyl ether.

10. a) Name and draw the structure of the monomers used for ATRP polymerization.

   b) Explain how ATRP is a living polymerization, i.e., give the mechanism for the extension of the lifetime of the free radical including chemical species involved in the synthesis.

   c) In the ATRP reaction the reaction mixture turned first blue, then green then back to blue. Explain these colour changes.

   d) Why was vitamin C added to the reaction mixture? Is this always necessary for an ATRP polymerization?

Answers

A

1. a	2. a	3. d	4. b	5. c	6. a	7. a	8. b	9. a	10. c
11. a	12. d	13. a	14. a	15. b	16. d	17. a	18. c	19. a	20. d
21. a	22. b	23. b	24. c	25. a	26. a	27. b	28. c	29. a	30. a
31. d	32. b	33. a	34. a	35. b	36. a	37. b	38. d	39. a	40. a
41. c	42. d	43. b	44. a	45. a	46. b	47. a	48. b	49. a	50. d
51. a	52. a	53. d	54. a	55. b	56. a	57. b	58. a	59. a	60. c
61. b	62. a	63. d	64. a	65. a	66. b	67. a	68. c	69. a	70. a
71. d	72. c	73. a	74. a	75. b	76. a				

Dissertation

The term dissertation carries different meaning according to the country's educational criteria. Basically a dissertation is an educational writing by undergraduate or postgraduate students. For the undergraduate (basically UGC CBCS system) student's dissertation they have to write about 60 pages on their given topic. The topics may be experimental or theoretical depending on their choice. In the experimental topics, students have to work in the laboratory according to the original research idea of the student /students' supervisor. On the basis of his result, he has to write a paper/project report. For theoretical topics, he has to go through the literature with the help of the supervisor on the selected topics. Student has to collect all the research works till date and write the summery by his own language. Also the student has to submit a digital presentation for the assessment.

Topic for dissertation on polymer

Polymer has many areas for dissertation including its synthesis, properties and application for both theoretical and experimental.

1. **Synthesis of polymer :** At first you have to choose specific polymer of your interest. From literature you have to find its methods of synthesis. If you want your dissertation theoretically then you have to summarize it and write it in your own language. If you want your dissertation experimentally then you have to think a new approach in any step throughout the synthesis method. The approach should be unique. There are lots of polymers with varieties of synthetic methods. So a large number of students can take this topic.

2. **Properties of polymer :** At first you have to choose specific polymer of your interest. From literature you have to find its properties of your interest. It may be glass transition temperature, mechanical properties, current-voltage properties, rheological properties, optical properties etc. If you want your dissertation theoretically then you have to summarize it and write it in your own language. If you want your dissertation experimentally then you have to change the processing techniques and check the properties variation. The approach should be unique. There are lots of polymers with varieties of properties. So a large number of students can take this topic.

3. **Applications of polymer :** This topic is harder than others. At first you have to choose specific polymer and its applications of your interest. From literature you have to find it. If you want your dissertation theoretically then

you have to summarize it and write it in your own language. If you want your dissertation experimentally then you have to find a new application or have to modify your material which gives better properties till now. The approach should be unique. There are lots of polymers with varieties of applications. So a large number of students can take this topic.

→ In each and every step student must ask help from the supervisor.

Printed by Libri Plureos GmbH in Hamburg,
Germany